CITIES AND COMPLEXITY

Cities & Planning Series

The Cities & Planning Series is designed to provide essential information and skills to students and practitioners involved in planning and public policy. We hope the series will encourage dialogue among professionals and academics on key urban planning and policy issues. Topics to be explored in the series may include growth management, economic development, housing, budgeting and finance for planners, environmental planning, GIS, small-town planning, community development, and community design.

Series Editors

Roger W. Caves, Graduate City Planning Program,
 San Diego State University
Robert J. Waste, Department of Public Policy and Administration,
 California State University, Sacramento
Margaret Wilder, School of Urban Affairs and Public Policy,
 University of Delaware

Advisory Board of Editors

Edward J. Blakely, *University of Southern California*
Robin Boyle, *Wayne State University*
Linda Dalton, *California Polytechnic State University, San Luis Obispo*
George Galster, *Wayne State University*
Eugene Grigsby, *University of California, Los Angeles*
W. Dennis Keating, *Cleveland State University*
Norman Krumholz, *Cleveland State University*
John Landis, *University of California, Berkeley*
Gary Pivo, *University of Washington*
Daniel Rich, *University of Delaware*
Catherine Ross, *Georgia Institute of Technology*

Karen Stromme Christensen

CITIES AND COMPLEXITY

Making Intergovernmental Decisions

Cities & Planning

SAGE Publications
International Educational and Professional Publisher
Thousand Oaks London New Delhi

Copyright © 1999 by Sage Publications, Inc.

All rights reserved. No part of this book may be reproduced or utilized in any form or by any means, electronic or mechanical, including photocopying, recording, or by any information storage and retrieval system, without permission in writing from the publisher.

For information address:

SAGE Publications, Inc.
2455 Teller Road
Thousand Oaks, California 91320
E-mail: order@sagepub.com

SAGE Publications Ltd.
6 Bonhill Street
London EC2A 4PU
United Kingdom

SAGE Publications India Pvt. Ltd.
M-32 Market
Greater Kailash I
New Delhi 110 048 India

Printed in the United States of America

Library of Congress Cataloging-in-Publication Data
Christensen, Karen Stromme.
   Cities and complexity: Making intergovernmental decisions / by Karen Stromme Christensen.
     p. cm.—(Cities & planning; v. 3)
   Includes bibliographical references and index.
   ISBN 0-7619-1164-2 (cloth: acid-free paper)
   ISBN 0-7619-1165-0 (pbk.: acid-free paper)
   1. Federal-city relations—United States. 2. State-local relations—United States. 3. Complex organizations—United States. 4. Professional employees in government—United States. I. Title. II. Series: Cities & planning series; v. 3.
JS348 .C57 1998
353.3'3'0973—ddc21   98-9069

99 00 01 02 03 04 05 7 6 5 4 3 2 1

Acquiring Editor:	Catherine Rossbach
Editorial Assistant:	Heidi Van Middlesworth
Production Editor:	Diana E. Axelsen
Production Assistant:	Nevair Kabakian
Typesetter/Designer:	Lynn Miyata
Indexer:	Trish Wittenstein

CONTENTS

Foreword	xi
Preface	xiii
Acknowledgements	xix
1. Planning in a Complex Intergovernmental System	**1**
Some Planning Dilemmas	2
The Complex Intergovernmental System, the Medium of Planning	2
Persistent Complexity and Delusions of Certainty	4
Confronting Actual Problem Conditions of Uncertainty and Complexity	8
2. Competing Theories of the U.S. Intergovernmental System	**11**
Intergovernmental System as Distinguished From Federalism, Intergovernmental Relations, and Intergovernmental Management	12
Theories of Intergovernmental Systems	13
Dual Federalism	14
Multicentered Federalism	15
Functional Federalism	16
Contradictions Among the Theories	17

3. **Interactions Among Governmental Institutions**   23
   Interdependence of Public Institutions   24
   Task Division Among Governments   27
   Interactions Among Governmental Agencies   32
      Collaboration Among Governments   32
         Autonomy   32
         Information Exchange   33
         Joint Learning   33
         Review and Comment   34
         Joint Planning   35
         Joint Funding   36
         Joint Action   37
         Joint Venture   37
         Merger   38
      Conflict Among Governments   39
         Autonomy   40
         Contradictory Information and Theory   41
         Negative Comments   41
         Competitive and Isolative Planning   41
         Prohibitory Legislation and Judicial Action   42
         Counter Programs and Funding   43
   Summary and Conclusions   43

4. **Dimensions Structuring the Intergovernmental System**   47
   The Vertical Dimension   48
   The Horizontal Dimension   52
   The Time Dimension   55
   The Decision Mode Dimension   56
      Majority   57
      Interaction in Deciding Processes   58
   Summary   62

5. **Sectoral Dynamics: Institutionalized Technological Elaboration and Effects on Area Interests**   65
   Sectoral Dynamics   66
      Sectors Distinct From Each Other   73
      Proliferation   73
   Significance of the Aggregate Pattern of Sectoral Dynamics   75

Sectoral Effects on Area Governments	76
Intended and Unintended Differential Effects	77
The Intergovernmental Grant-in-Aid System	79
Territorial Treatment of Sector Policy	79
Disjointedness of Multiple Sectors Affecting Area Governments	80
Area Interests Suppressed by Sectors' Dominance	82
Significance of the Pattern of Sectoral Dominance Over Area Interests in the Aggregate	84
Deterred Debate Over Goals	84
Skewed Democratic Access	84
Systemic Piecemeal Decision Making Without Self-Correcting Feedback Loops	85
Conclusions	85
6. Delusions of Certainty and Their Consequences for Expectations of Government	**89**
Variable Problem Conditions With Respect to Expectations for Government	90
Problem Conditions and Associated Expectations	90
Known Technology, Agreed Goal	91
Unknown Technology, Agreed Goal	92
Known Technology, No Agreed Goal	94
Unknown Technology, No Agreed Goal	95
Conclusions Regarding Variable Problem Conditions	95
Governmental Predispositions Toward Certainty	97
Administrative	98
Legal	99
Political	100
Behavior Arising From Delusions of Certainty	100
Intergovernmental System Dynamics and Delusions of Certainty	103
Programming and Stable Sectors	103
Experimentation and Sectoral Interaction	104
Bargaining and Sectoral Effect on Area Interests	106
Confusion Stemming From Prematurely Programmed Sectoral Effects on Diverse Area Conditions and Preferences	109
Conclusions	112

7. **Conducting Public Policy in Conditions of Uncertainty**   115
   A Strategy of Variability   116
   Variable Expectations of Government   118
   Variable Forms of Policy   119
      A Framework of Variable Policy Form   120
         Prespecified Process and Prespecified Outcome   121
         Open Process and Prespecified Outcome   122
         Prespecified Process and Open Outcome   125
         Open Process and Open Outcome   129
      Adopting Appropriate Policy Form   130
   Variable Forms of Organization   132
      Stable Rules for Known Technology and Agreed Goal   132
      Change and Expansion for Unknown Technology and
         Agreed Goal   133
      Agreed Rules of the Game for No Agreed Goal
         or Outcome   134
      Redundant Complexity for Unknown Technology and
         No Agreed Goal   134
   Variable Forms of Planning   136
      Planning for Known Technology, Agreed Goal   136
      Planning for Unknown Technology, Agreed Goal   137
      Planning for Known Technology, No Agreed Goal   140
      Planning for Unknown Technology, Unknown Goal   141
      Summary of Planning With Respect to Variable Problem
         Conditions   143
      Intergovernmental System Influences on Planning Practice   143
      Challenging System Constraints on Planning Practice   145
   Summary   148

8. **Conclusions for the Intergovernmental System**   151
   Summary of the Argument   152
      Sectoral Versus Area Governments   152
      Certainty Versus Uncertainty   152
   Contradictions of Federalism Today   153
      Current Intergovernmental System Considered in Terms of
         Three Prevailing Models of Federalism   154
      Enduring Principles of Federalism and Current Operations   155
      Federalism Is Multiple, Not Unitary   156

Federalism Is Based on Checks   156
　　　Federalism Is Conservative   157
　　　Conclusions: The Territorial Basis of Federalism Is Weak; Long
　　　　Live Federalism   158
　　　Policy Implications: Toward a New Policy Debate and
　　　　New Policies   158

9. **Conclusions for Decision-Making Practice**   161

　　　References   167

　　　Index   175

　　　About the Author   177

FOREWORD

The study of cities is a dynamic, multifaceted area of inquiry that combines a number of disciplines, perspectives, time periods, and actors. Urbanists alternate between examining one issue through the lens of a single discipline and looking at the same issue through the lens of a number of disciplines to arrive at a holistic view of cities and urban issues. The books in this series look at cities from a multidisciplinary perspective, affording students and practitioners a better understanding of the multiplicity of issues facing planning and cities and the emerging policies and techniques aimed at addressing those issues. The series focuses on both traditional planning topics such as economic development, management, and control of growth, and geographic information systems. It also includes broader treatments of conceptual issues embedded in urban policy and planning theory.

The impetus for the *Cities & Planning* series originated in our reaction to a common recurring event—the ritual selection of course textbooks. Although we all routinely select textbooks for our classes, many of us are never completely satisfied with the offerings. Our dissatisfaction stems from the fact that most books are written for either an academic or a practitioner audience. Moreover, on occasion, it appears as if this gap continues to widen. We wanted to develop a multidisciplinary series of manuscripts that would bridge the gap between academia and professional practice. The books are designed to provide valuable information to students and instructors and to practitioners by going beyond the narrow confines of traditional disciplinary boundaries to offer insights into the urban field.

Karen Stromme Christensen's *Cities and Complexity: Making Intergovernmental Decisions* is a major contribution to understanding the modern city. In *Cities and Complexity,* Christensen provides an innovative and promising "map" to guide students, professors, and practitioners through the field of the urban world—a world typified more often than not by complexity and uncertainty within a given metropolitan region, and compounded at higher levels by increased complexity and uncertainty when one attempts to understand cities with the complex web of national and international intergovernmental and economic systems. Christensen provides order, meaning, and predictability to urban intergovernmental relations/complexity landscape by reducing the complexity along an axis corresponding to either specialization or territory. Her resulting matrix promises to help urbanists better understand complex issues and practitioners better expand government's capacity to address uncertainty. *Cities and Complexity* will represent a major contribution to both urban theory and urban practice.

—Roger W. Caves
San Diego State University

—Robert J. Waste
California State University, Sacramento

—Margaret Wilder
University of Delaware

PREFACE

This book emerged from my frustrations working at the U.S. Department of Housing and Urban Development (HUD) during its heyday. We were full of innovation and ambition to solve pressing urban problems. I pursued my interagency and intergovernmental projects with enthusiasm, but faced countless difficulties.

Others shared my disappointment as confidence in government plummeted. Today, politicians, planners, administrators, and concerned citizens are frustrated by the confusion and tenacity of a government they feel unable to direct. Only 20% of Americans "trust the government in Washington to do what is right" most of the time, down from 73% in 1958 (Ladd, 1995). Expectations of efficiency and effectiveness often go unmet. Beyond this thwarted political urge to manage government lurks despair over its apparent inability to solve problems. This frustration leads to blaming government and trying to impose simplicity and certainty.

Yet the actual conditions in today's government are complex, uncertain, and misunderstood. Government contains thousands of agencies at multiple levels interacting in seeming chaos. Government's goals and task technologies tend to be ambiguous and uncertain. Denying complexity and uncertainty or ordering them away by simple edict tends to induce harmful side effects, prompting further rounds of reforms and further complications.

I left HUD for the University of California, Berkeley, to address the following questions: How does the intergovernmental system work? Why is interagency cooperation hard to achieve? How can planners be effective when the system they work in is so complex and chaotic? I explored these questions as I earned my master's and PhD degrees and joined the faculty. I hope the answers I present here will enable analysis to move beyond current frustrations and dilemmas to new solutions.

Years of teaching and research helped me develop a new theory of the structure and dynamics of the U.S. intergovernmental system, summarized as follows.

Interactions among governmental units create the immensely varied world of government through numerous combinations. They operate according to two bases of organization: specialization and territory.

Most often, intergovernmental activities occur among agencies working in the same functional specialization, such as housing, water quality, or transportation. Agencies working in the same specialization share the same general mission (e.g., more and better transportation) and often the same profession (e.g., transportation engineering). They perform interdependent tasks addressing their mission. Their shared mission, profession, and tasks link the agencies together vertically, up and down through federal, state, and local governments.

U.S. government is also organized by territory—that is, spatially by cities, counties, and states. In contrast to the vertically linked specializations, territory-based government can be seen as horizontal.

When specializations' programs hit the ground in a territorial government, they tend to be treated by the local agency in that specialization. For example, housing programs tend to be treated by the local office of housing and community development or the local housing authority. Because this happens in specialization after specialization, from water quality to mental health, specializations dominate decision making.

Over time, specializations reformulate their programs—centralizing, decentralizing, privatizing, and so on—according to the interplay of politics and technology development in the specialization. Each new version is applied to many places, often generating surprising side

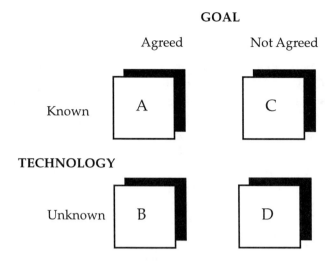

Figure I.1. Prototype Conditions of Public Problems

effects that prompt further reformulations and complications. Thus, programs grow in number and complexity even though policymakers strive for simplicity.

Collectively, many specialized programs affect a territory in ways that are necessarily disjointed. For example, a city suffers funding recision for its revitalization program while it decentralizes health services to neighborhoods and a week later receives a new jobs development program. The city is left to accept, adapt, or resist each specialization's program. The specializations' dominance and expectations of certainty may harm real people and places and skew democratic access and curb debate over goals.

The presumption of certainty contributes to these perverse intergovernmental dynamics. To help expand government's capacity to address uncertainty, I have adapted Figure I.1 (Thompson & Tuden, 1959) for policy and planning.

Figure I.1 differentiates government problems according to their types of uncertainty. Recognizing that goals may be agreed on or not and that technology (the instrumental method for accomplishing a government task) may be known (proven effective) or not yields four distinctly different problem conditions. Box A is agreed goal, known technology; Box B is agreed goal, unknown technology; Box C is no agreed goal, known technology; Box D is no agreed goal, unknown technology.

Government performance should vary in response to these different problem conditions. This book presents forms of planning, organization, and policy design addressing different conditions of certainty and uncertainty. Acknowledging and addressing different problem conditions with appropriately different government performance offers a fresh way around current frustrations and self-defeating reforms.

For example, when people agree on the goal (e.g., welfare reform) but have no proven effective ways of achieving the goal (e.g., helping welfare recipients obtain steady, decent employment), the situation calls for discovering or inventing effective new ways. To do so calls for innovation and experimentation and corresponding types of planning and policy (e.g., a research and development program). Other problem conditions call for different government performance and, accordingly, different kinds of planning and policy.

Similarly, this framework can help select government "reinventions" (policy designs) appropriate for particular problem conditions. It offers a lens through which to examine ideas that seem attractive but unquestioned. For example, privatizing government work, such as providing housing vouchers, makes sense where a market functions fairly well and different households make different choices (known technology but no single, agreed goal). In contrast, privatizing government makes no sense for banning toxics (known technology, agreed goal). This book sets forth strategies that encourage the invention of still more specific policies, plans, organizational designs, and everyday practices tailored to particular problem conditions and political preferences.

The book develops the intergovernmental theory and contingent approaches to uncertainty as follows. Chapter 1, "Planning in a Complex Intergovernmental System," poses the question, How does the intergovernmental system shape planning? Chapter 2, "Competing Theories of the U.S. Intergovernmental System," examines competing theories about the intergovernmental system's structure. Chapter 3, "Interactions Among Governmental Institutions," describes current interactions among governmental units. Drawing from theories and actual examples, Chapter 4, "Dimensions Structuring the Intergovernmental System," extracts the U.S. intergovernmental system's key dimensions: function (vertical), territory (horizontal), and time. Chapter 5, "Sectoral Dynamics: Institutionalized Technological Elaboration and Effects on Area Interests," analyzes intergovernmental dynamics along the functional dimension, at the juncture of function and territory, and how many different functions affect an area over time. Chapter 6, "Delusions of Certainty and Their Consequences for Expectations of Government," examines how government's failure to meet expectations stems in part from the incompatibility between its expectations of certainty and the essential uncertainty characterizing its tasks, and offers a matrix to address uncertainty. Chapter 7, "Conducting Public Policy in Conditions of Uncertainty," proposes contingent ways of planning and policy design to address uncertainty. Chapter 8, "Conclusions for the Intergovernmental System," draws conclusions for the intergovernmental system. Chapter 9, "Conclusions for Decision-Making Practice," recommends everyday practices to facilitate adopting contingent approaches to uncertainty.

Beyond my theoretical contributions, I hope my efforts to understand and work more effectively in the intergovernmental system will help students, planners, and policymakers do the same. As public decision makers become more effective and responsive, I hope they will help renew public confidence in government and thereby increase its capability to address uncertainty in the future.

To Michael

ACKNOWLEDGMENTS

First, special gratitude to my academic mentors: Melvin Webber, Judith Innes, Martin Landau, and Michael Teitz. Their guidance has been invaluable; their criticism has always been constructive, and their support has always felt unlimited and has continued over many years.

I would also like to thank other professors and colleagues who helped me during different phases in the development of this book: John Bryson, Frederick Collignon, Vicki Elmer, Darla Guenzler, Victor Jones, Neema Kudva, Ann Markusen, Tak Nakamoto, Richard Phillips, Craig Thomas, Robert Thompson, Aaron Wildavsky, Nadine Wilmot, and David Wilmoth. I particularly wish to thank the writing group: Judith Gruber, Judith Innes, Katherine Roper, Christine Rosen, and Eleanor Swift. I am grateful to Sharon Beaty and Wayne Heiser, who provided perceptive and useful student reaction. Sincere thanks to the supportive and encouraging editors: Roger Caves, Catherine Rossbach, Robert Waste, and Margaret Wilder. Even with all this wonderful help, the responsibility for errors rests squarely with me.

In addition, I am intellectually indebted to Robert Biller, Jeffrey Pfeffer, and Horst Rittel, who have all influenced my thinking profoundly, though none has had direct involvement in this effort.

If I am not being too presumptuous, I would also like to thank James D. Thompson, whom I never met but whose insights spurred me. I have tried to follow his instruction "to study organizations in toto, and, for that purpose, the significance of the . . . system approach and the certainty/uncertainty dimension" (Thompson, 1967, p. 163).

My deep gratitude to Christine Amado, Kate Blood, Norma Montgomery, and Theresa Ojeda at Berkeley's Institute of Urban and Regional Development, who typed and processed several versions and revisions with commendable skill, care, and patience.

Finally, I would like to thank my husband, Boake, for all his support and encouragement. I wish he could have lived to see this in print. This book is dedicated to our son.

CHAPTER 1

PLANNING IN A COMPLEX INTERGOVERNMENTAL SYSTEM

Planners and managers hate uncertainty as much as most people, and they spend their working lives trying to reduce it on a public scale into the problematic future. The complex intergovernmental system through which they act constrains them, however, and compounds their difficulties. This book tries to make their work easier by showing how the U.S. intergovernmental system functions.

This introductory chapter describes how planning is enmeshed in complexity and uncertainty. It begins with dilemmas inherent in planning and depicts the complex intergovernmental system as the medium of planning. Next, it discusses the contradiction between the assumptions and expectations of planning and the intergovernmental system through which planning takes place. Then, it turns to the persistent delusions of certainty that planners, managers, and politicians have in the face of the intergovernmental system's complexity. The chapter concludes by proposing some benefits from responding to actual uncertainty and complexity.

SOME PLANNING DILEMMAS

Planning is fraught with dilemmas. *Planning* is a type of public decision making.[1] It is defined here as a deliberative process of devising a set of actions to change the future course of events for some public purpose. The dictionary's definition is crisper: "to devise a course of action" (Webster-Merriam, 1956, p. 644). The extended definition used here emphasizes the future, change, deliberative processes, and value-laden public purpose—all of which start to expose dilemmas inherent in planning.

Planning quandaries include, How can we know the future? What is the public interest? How can we know which values are right? What is the relationship between means and ends? This book particularly addresses the mismatch between the assumptions of planning and management and the actual operations of the intergovernmental system. In doing so, it also addresses how the intergovernmental system shapes the answers to these planning dilemmas.

THE COMPLEX INTERGOVERNMENTAL SYSTEM, THE MEDIUM OF PLANNING

The intergovernmental system provides the institutional context of planning. The intergovernmental context of planning has always been important since planning began in the United States around the turn of the 20th century. The federal[2] government promoted planning through the Hoover Commission, the National Planning Board, and urban renewal and the subsequent war on poverty intergovernmental planning and programs; later with environmental issues; and recently as more state and regional intergovernmental planning has emerged. Thus, the intergovernmental system historically has been integral to planning.

Moreover, the importance of the intergovernmental system for planning has continuously grown because the system is developing in two ways. First, it is proliferating in sheer numbers of agencies, programs, and planning activities. Second, individual activities are becoming more intertwined with interdependent actions by multiple levels and, accordingly, are being understood as intergovernmentalized. Thus, the complexity of the intergovernmental system over time has become an increasingly significant factor in planning.

Despite its importance, the intergovernmental system is so complex that it seems to defy understanding. Why are relations among governments so confusing? Beyond this intellectual cry for comprehension lie two practical concerns. First, how can planners, managers, politicians, and the general public be effective in realizing their demands of government when agencies are so numerous and the relationships of their responsibilities are so complex? Second, how can policymakers be effective in arranging new intergovernmental solutions to diverse problems when even expert professionals cannot see through the chaos?

> **Box 1.1**
>
> Why does the United States seem to require such complexity to accomplish tasks that are relatively simple? For example, nine sources of public subsidy—private lending, a nonprofit bridge loan, a nonprofit housing agency, a city agency, a state agency, a federal agency, a city housing element, a regional fair share plan, and a private for-profit housing construction agency—contribute to the production of a 27-unit moderate-income housing project (Pamuk & Christensen, 1989). This example sounds like a joke (How many U.S. agencies does it take to screw in a lightbulb?), but it is accurate. This order of complexity mystifies planners as well as the public.

Although the intergovernmental system seems incomprehensible, planners cannot ignore it because it shapes planning. Decisions are reached through intergovernmental deliberation, and implementation demands intergovernmental action. No level of government is autonomous; all are interdependent.

On the one hand, each jurisdiction addresses its own individual goals through individual planning for itself. But, because the jurisdiction or agency is neither autonomous nor self-sufficient, its internal planning is necessarily partial. What actually happens to its planned project depends on actions outside the jurisdiction. This interdependence occurs throughout the system, from the White House to the town of

Albany, California, from the Environmental Protection Agency to the Child Care Council of Alameda County. For example, a single local economic development agency, trying to lure a shopping center to locate in its jurisdiction, must deal with competition from neighboring cities, the metropolitan transportation commission, the state department of transportation (to secure a new freeway off-ramp), and the state environmental impact assessment process, which in turn involves many other actors. Pervasive intergovernmental interdependence means that even planning for a single public entity is laden with complexity, confusion, and uncertainty.

On the other hand, jurisdictions also participate in intergovernmental planning. Sometimes the intergovernmental deliberation occurs through partisan mutual adjustment (Lindblom, 1965). A jurisdiction does something; another jurisdiction reacts; the first adjusts; and so on. Sometimes jurisdictions engage in explicit intergovernmental planning. Planners and policymakers often complain of problems in coordination and conflict, of confusing duplication, overlap, competing mandates, unnecessary meddling, too many demands, too narrow perspectives, and inefficiency.

Because of this interdependence, the intergovernmental system constitutes more than the environment of planning; it forms the medium of planning. Used in this way, *medium* means "a substance through which a force acts or an effect is transmitted" and "that through or by which anything is accomplished" (Webster-Merriam, 1956, p. 523).

The intergovernmental system shapes the content and direction of planning by framing how goals are formulated, what gets paid attention to and what does not get paid attention to, how technologies are developed, which technologies are developed or not, what gets acted on and not acted on, what falls through the cracks, and how policies, plans, and programs change.

PERSISTENT COMPLEXITY AND DELUSIONS OF CERTAINTY

Part of the reason the intergovernmental system is difficult to direct lies in its original design to protect against tyranny or, more loosely, direction. The original tyranny-proof, localist, and fractionated design was

Planning in an Intergovernmental System

also sufficiently unspecified, flexible, and open that it could respond to the demands of an ever-developing and increasingly interdependent political economy. Partly because the original governmental design was simultaneously both adaptable and resistant to major structural change, the system of government never underwent a formal reorganization. Instead, it evolved over 200 years through a variety of piecemeal, unplanned adaptations that collectively resulted in today's complex intergovernmental system.

Another part of the difficulty in trying to direct government lies in the contradiction between the intergovernmental medium of planning and public expectations of efficiency and effectiveness. Planning aims at ensuring future certainty in a complex, dynamic intergovernmental system that is rife with turbulence and uncertainty. Thus, what is planned eventually becomes changed through the intergovernmental medium. The results of planning fail to meet intentions. These outcomes are not only disappointing, they seem confusing, as if the system is unmanageable.

The incompatibility between expectations of planning and the intergovernmental system is compounded by each side's persistence. Regardless of reality, planners, managers, and politicians persist in their efforts to impose certainty. Alternatively, the issue may be seen as governments' pervasive interdependence and proliferation. Why does the intergovernmental system persist in its complexity, when public officials keep trying to simplify it?

Evidence of the public's desire for control abounds throughout the last century (Chisholm, 1989) and in today's newspapers. Why do governmental institutions proliferate despite periodic attempts to streamline, simplify, and reduce government? This question seems to defy conventional political analysis because complaints about the "bureaucracy's" size, ineptness, and (paradoxically) monolithic power seem to ignore party lines. All recent presidents since Johnson have pledged and attempted to disrupt government's pattern of growth, including the current attempt to "reinvent government."

As Table 1.1 shows, government programs continue to grow. Even during the 1980s and 1990s, when administrations tried to reduce programs and consolidate them in block grants, the number of programs increased.

TABLE 1.1 Federal Categorical and Block Grants, Selected Fiscal Years 1975-1995

Number and Outlays (current dollars in billions)

	1975 No./$	1978 No./$	1981 No./$	1984 No./$	1987 No./$	1989 No./$	1991 No./$	1993 No./$	1995 No./$
Block	5/4.6	5/11.5	6/10.0	12/13.0	13/13.1	14/12.7	14/16.4	15/20.5	15/22.8
Categorical	442/38.2	492/56.8	534/77.9	392/77.8	422/93.2	478/106.9	543/133.4	578/171.1	618/202.9
Total	447/42.8	497/68.3	540/87.9	404/90.8	435/106.3	492/119.6	557/149.8	593/191.6	633/225.7

Source: Advisory Commission on Intergovernmental Relations (1995), based on Catalogue of Federal Domestic Assistance, U.S. code, and federal agency contacts.

Uncontrolled growth and weak accountability are symptoms of untrammeled specialization. In addition to attempts to reduce government, efforts to respond to these problems include (1) frequent political promises of "comprehensive" packages, whether for cities, the economy, or other perceived wholes; (2) enthusiastic but premature adoption of whole system technical applications, such as simulation models, program, planning, and budgeting systems, zero-based budgeting, and related management schemes, such as management by objectives and total quality management; (3) development and growth of generalist, overarching units such as the Office of Management and Budget, the Congressional Budget Office, and equivalents at state and local levels; and (4) metropolitan councils of government and numerous coordinating committees and task forces. All are but pleas for integrating[3] an increasingly fragmenting, specializing system.

Planners and managers have difficulty understanding the intergovernmental system. Consequently, they try to cope with their immediate situation. They respond not by looking to the larger intergovernmental system, but rather by searching through their repertoire of familiar responses for something to solve the immediate problem (Cyert & March, 1963). The results of not understanding the system include frustration, confusion, and continued efforts to plan better and coordinate better[4] to try to impose certainty. Unable to see the contradiction, planners and managers engage in self-defeating attempts to simplify reality to make it fit expectations. They tend to assume (1) the certainty of means and ends; and (2) that the agencies they are working through act with certainty as well.

Actually, the norms of individual governmental agencies reinforce the contradiction. In contrast to the complex intergovernmental system they compose, many of the individual governmental agencies are modeled on bureaucratic (Gerth & Mills, 1946) certainty. For legal, political, administrative, and social (Powell & DiMaggio, 1991, p. 27) reasons, public agencies seem bound to create this veneer of certainty and aim for associated bureaucratic norms of predictability, equity, accountability, efficiency, and effectiveness. Yet, the problems governmental agencies address are rarely so certainly solved as to be frozen answers to political questions.

The mismatch between expectations and actual conditions in the intergovernmental system is not being addressed. Part of the reason lies in the reciprocal relationship between the intergovernmental system and planning. The intergovernmental system shapes planning, whereas planning works on and through the intergovernmental system. Planning efforts to control become self-defeating and perpetuate the mismatch between expectations and actual conditions. Furthermore, the anomaly of proliferating specialized governmental agencies in spite of wide support to reduce them serves latent functions (Merton, 1957), such as learning, adaptation to diversifying public demands, and narrowing of purpose. In effect, both complexity and simple certainty are desired.

CONFRONTING ACTUAL PROBLEM CONDITIONS OF UNCERTAINTY AND COMPLEXITY

When planners do not understand how the intergovernmental system works, their plans are liable to be distorted by the system or bypassed altogether. Planning that mistakenly assumes certainty and simplicity can become perverse by generating unanticipated results and allowing errors to continue and to affect more people and places than necessary. Thus, planners engage in counterproductive behavior that perpetuates the uncertainty. In doing so, planners limit themselves and limit the role of planning in society.

When planners understand the intergovernmental system they are working in, they can address its complexity and uncertainty directly and then be both more effective and more efficient in addressing their goals. In the course of doing so, they can reduce complexity and uncertainty.

This chapter set forth a central question: How does the intergovernmental system act as the medium for planning and shape planning outcomes? The rest of the book addresses this question by developing an understanding of (1) how the intergovernmental system works, (2) what effects the system has on planning and policy, and (3) how to plan and manage more effectively.

NOTES

1. Despite the following distinctions, much of what is termed here as *planning* applies to decision making in the public arena. Planning can be understood as a subset—a thoughtful version—of decision making. As used in this book, planning is public and

Planning in an Intergovernmental System

deliberative, whereas some decision making may be private and undeliberative, that is, unreflective, perhaps impulsive, reactive, or simply following routine. Planning is goal seeking or problem solving, whereas some decision making may be so situation specific and ad hoc that it does not serve any further purpose. Planning addresses the future and aims at anticipating consequences in advance of action, whereas some decision making may not speculate or consider consequences.

2. In the United States, the national government is conventionally known as the "federal government." It will be referred to that way in this book, even though it would be correct to refer to it as the "national government in the federal system of governments."

3. Frustrated by a seemingly intractable system, some planners and policymakers have looked to larger-scale coordination and control. Despite such attempts, most citizens, politicians, and academics fear any alternative to a fractionated, conflicting, and confusing intergovernmental system because they believe central control is the only antidote.

4. "If only we can find the right formula for coordination, we can reconcile the irreconcilable, harmonize competing and wholly divergent interests, overcome irrationalities in our governments structures" (Seidman, 1970, quoted in Chisholm, 1989, p. 1).

CHAPTER 2

COMPETING THEORIES OF THE U.S. INTERGOVERNMENTAL SYSTEM

The government of the United States is not a unitary government but rather myriad units that constitute an intergovernmental system. Its basic elements are the multiple governmental organizations—special purpose, general purpose, legislative, judicial, administrative, planning, and implementing—that extend from the national to the local level. Like chemical elements, these governmental units compose the immensely varied, complex world of governance through their numerous combinations and interactions.

This chapter sets out broad, prevailing theories about these interactions. The theories impose a conceptual order on the apparent chaos of government and direct attention to basic dimensions structuring intergovernmental relations. After some brief distinctions, the chapter describes three dominant, competing theories of the intergovernmental system. Then it discusses the contradictions among them. By exploring competing theories, discussion debunks some conventional views, challenges others, and begins to consider the contemporary intergovernmental system.

INTERGOVERNMENTAL SYSTEM AS DISTINGUISHED FROM FEDERALISM, INTERGOVERNMENTAL RELATIONS, AND INTERGOVERNMENTAL MANAGEMENT

This book focuses on the medium of planning, the intergovernmental system. A system is "an assemblage of objects united by some form of regular interaction or interdependence" (Webster-Merriam, 1956, p. 863). The intergovernmental system is both related to and distinguished from federalism, intergovernmental relations, and intergovernmental management, as briefly follows.[1]

Federalism concerns the division of powers between states and the federal (national) government. (In contrast, the intergovernmental system also includes local governments and many special purpose agencies.) Although the federalism literature tends to be academic, issues of federalism arise periodically in public policy through the courts and the rhetoric of political change. For example, politicians may promote programs under the rubric of new federalism, or argue that the founding fathers meant to have states make decisions on welfare. For purposes here, federalism refers to the basic structure of the intergovernmental system.

The term *intergovernmental relations* originated in the 1930s and was associated with service delivery. It recognizes the complexity and interdependence of the intergovernmental system. Wright (1988) traces the development of intergovernmental relations, which work through mechanisms of intergovernmental finance and program design. For purposes here, intergovernmental relations refers to interactive activities between governments within the intergovernmental system.

Intergovernmental management is distinguished from intergovernmental relations mainly by being goal directed. Intergovernmental management is characterized by its (1) problem solving; (2) means of coping with the intergovernmental system as it is, including perspectives on how and why interjurisdictional changes occur; and (3) emphasis on communication networks (Agranoff & Rinkle, 1986, p. 5; Wright 1983). The literature tends to be both descriptive and normative. For purposes here, intergovernmental management refers to particular cases of and ways of improving planning and problem solving in the intergovernmental system.

In sum, federalism refers to multigovernment structure; intergovernmental relations refers to interactive activities between governments; and intergovernmental management refers to purposive, interactive activities between governments. In contrast, the intergovernmental system refers to patterns and principles that govern the dynamic relations among the interdependent governments. Thus, the intergovernmental system includes federalism, intergovernmental relations, and intergovernmental management.

Understanding the intergovernmental system can contribute to thinking about federalism, intergovernmental relations, and intergovernmental management. Thus, the system perspective helps to clarify the federalist structure in a dynamic world. At the same time, understanding the patterns of intergovernmental processes helps provide a basis for interpreting the constant flux of intergovernmental activities. Finally, understanding the patterns and governing principles of intergovernmental processes can help planners and managers be more effective in their problem solving.

THEORIES OF INTERGOVERNMENTAL SYSTEMS

Federalism has four properties: (1) two levels of government rule; (2) the same territorially based people; (3) each level has at least one area of action in which it is autonomous; and (4) there is some guarantee of the autonomy of each government in its sphere (Riker, 1964). Most scholars and public servants would accept this or a similar bare-bones definition of federalism as an organizational concept.

How federalism behaves, however, can be controversial. Of the hundreds of interpretations of federalism's meaning and operation, three theories are dominant. These three theories (Figure 2.1) are the traditional dual federalism, pictured as a layer cake; multicentered federalism, pictured as a marble cake; and functional federalism,

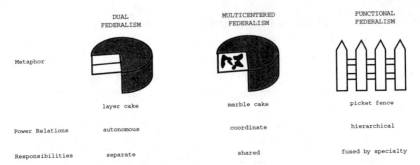

Figure 2.1. Three Prevailing Theories of Federalism

pictured as a picket fence (Sanford, 1967).[2] Each is described briefly before their contradictions are examined.

Dual Federalism

Dual federalism's (Corwin, 1934, pp. 47-48) core characteristic is the separation of powers, authorities, and functions of the two governmental levels. The top layer of the cake represents the national level, the lower layer represents the state level. Because their responsibilities are entirely distinct, the two levels of government neither collaborate nor direct each other; they are autonomous.

This separation of powers and the two levels' autonomy were very likely intended by the U.S. Constitution. In 1787, government tasks were seen as few, simple, and easily divided. For example, the national government would mint coins; the state governments would regulate alcohol. Thus, the theory of dual federalism seems congruent with the founding era, both in concept and in practice.

Furthermore, Thomas Jefferson's writings (Huntington, 1959) clearly stress territorial division of function. Even Madison (who is often read as the founder of dynamic pluralism in the federalist principle) said, "The federal constitution forms a happy combination in this respect; the great and aggregate interests being referred to the national, the local and particular to the state legislatures" (Madison, Hamilton, & Jay, 1937/1787-1788, p. 60). Significantly, all the founding fathers were convinced that the general principle of separation of powers was the essential insurance against government tyranny, their principal concern.

Multicentered Federalism

In a challenge to this traditional theory, Morton Grodzins (1966) posits that governmental responsibilities are not separate, as in a layer cake, but rather shared, as in a marble cake (McLean, 1952).[3] The chocolate flavor swirls back and forth, mixing with the vanilla flavor. The two levels are not distinct, but combined in a complex, random pattern. Because the theory was developed to explain the messy complexity of reality, it became known as multicentered federalism.

Instead of separate levels of government, so distinct in their functions that their relations are nearly static, the multicentered theory sees the levels mingling in a variety of changing tasks so that relations are dynamic. It is called "multicentered" to convey how decisions emerge from many points and from interactive processes.

> In this view, closely related to the various theories of social contract, [federalism] is characterized by the desire to build society on the basis of coordinative rather than subordinative relationships and by the emphasis on partnership among parties with equal claims to legitimacy. (Boehm, 1931, in Elazar, 1968, p. 354)

Historical analysis (Boehm, 1931, in Elazar, 1968) suggests that the origins of federalism lie in the effort to preserve diversity and at the same time promote union. The result is an exceedingly complex and variable division of authority. This multicentered, shared responsibility is incompatible with conventional models of traditional, bureaucratic organization culminating in a single, ultimate authority. Thus,

> federalism [is] fundamentally different from centralization and decentralization. There are mechanisms of partisan mutual adjustment, cooperation, and conflict resolution for noncentral coordination of relationships in federal systems that cannot be characterized as centralization and decentralization. (Ostrom, 1977)

"Federalism requires mutuality, not hierarchy, multiple rather than single causation, a sharing instead of a monopoly of power" (Wildavsky, in Ostrom, 1977). Neither level can direct or coerce the other. Public choices emerge from the levels' interaction.

Recognizing that policy emerges from interaction among governmental agencies directs attention to how they interact. Key issues include what prompts their interactions (e.g., federal grants in aid), motivations, and behavior. Are the interactions harmonious or competitive (e.g., Kincaid, 1991)? Are they cooperative (e.g., Chisholm, 1989; Elazar, 1991)? Are they redundant (e.g., Bendor, 1985; Landau, 1973)? In turn, those concerned with intergovernmental management ask about conditions that encourage certain kinds of cooperative problem-solving behavior (e.g., Agranoff & Rinkle, 1986) and how intergovernmental arrangements fit with policy (O'Toole, 1993).

In sum, the multicentered theory of federalism emphasizes coordinate rather than subordinate, multicentered rather than single-centered, and shared rather than independent government, perpetuated by dynamic interaction rather than static separation.

Functional Federalism

The third dominant theory of federalism is sometimes termed *nominal federalism* to convey that it is not truly federalism, but federalism in name only. Functional federalism asserts that neither autonomy nor coordinate sharing among government levels accurately reflects current reality. Instead, levels of government are related hierarchically. A picket fence provides the customary metaphor for functional federalism (Sanford, 1967, p. 80; Wright, 1978). The pickets represent numerous governmental functions, which cross traditional territorial layers of government. As they link each functional specialty, these pickets dominate. The thin rails represent the weak territorial levels.

In functional federalism, governments are ordered, one superior to another. Imagine a gigantic organization chart with thousands of cities and counties at the bottom, channeled up to 50 states, in turn channeled up to the apex, a single national government. This image of corporation and integration captures the idea of nominal, or centralized, federalism and emphasizes its compatibility with advanced capitalism.

Regardless of initial design, the intergovernmental system has evolved into centralized federalism. Four perspectives contribute to this view (Wright, 1978): decision making by a power elite (Mills, 1956, p. 6), benefiting corporate enterprise (Advisory Commission on Intergovernmental Relations, 1961), serving special interests (Lowi, 1969; McConnell, 1966), and state and local governments serving as national government's administrative districts (Derthick, 1970).

Competing Theories

The political economy is vertically integrated,[4] with sets of interlocking institutions linking spheres, both public and private, from local to national levels within each functional specialty (Lowi, 1969; McConnell, 1966). Government structure, all but inseparable from the larger economy, takes this vertical form. Thus, local businesses, supported by local interest associations, are linked to statewide, regional, and national businesses, supported by their affiliated interest associations. Their specialties, or functions, are supported by respective government operations at each level.

An elaborate grant-in-aid system, buttressed by regulatory mechanisms, links federal, state, area, county, and city governments by function. For example, the U.S. Department of Agriculture, working through its local extension service as well as through state agencies and various price-supporting regulatory mechanisms, serves the agriculture industry. In most functional areas, such as housing and transportation, private industry reinforces the vertical pattern through similarly connected associations. Specialization serves as the organizing principle, linking private interests and government levels across traditional boundaries in service of their specialty.

CONTRADICTIONS AMONG THE THEORIES

The three prevailing theories describe distinctly different relations of power and responsibilities among the governments. Dual federalism posits autonomy, multicentered federalism posits sharing, and functional federalism posits hierarchy (Wright, 1988). Although the theories are incompatible, all are considered valid.

If dual federalism was intended, it was hardly realized. National and state governments have always shared actual responsibilities (Elazar, 1962; Grodzins, 1966; Kestnbaum, 1955[5]), for example, for waterways. And as early as 1819 in *McCulloch v. Maryland*, the Supreme Court deemed the intermingling constitutional.

Nonetheless, dual federalism retains a crucial legitimacy in its ideology. "Jefferson furnished the ideology and semantics for division of power in America; Madison described the reality" (Huntington, 1959, p. 197). At the time the constitution was framed, the small communities of propertied men really could permit true representation and direct response to locally perceived needs; these communities really did constitute discrete, autonomous interests. Antifederalists could argue (Lewis, 1967) that different types of people require different laws, which they must set for themselves in small communities.

Thus, localism, home rule, and democracy came together, seemingly synonymous. Furthermore, many of these communities afforded everyone (excluding slaves, children, and women) full, active participation in governmental decisions. From de Tocqueville's view, this localist form of government gave citizens the capacity to build, to act, to become government, rather than merely to resist central authority (Hoffman, 1959). Both the inspiration and the ideology of self-government seemed embedded in local government's separation and virtual independence from national government. The doctrine of division of powers is still invoked whenever states or localities oppose a national policy. Cries for "states' rights" and "home rule" and complaints about national "meddling" all testify to the continuing force of the ideology of separate, independent governments regardless of the evidence of shared powers.

Although multicentered federalism is more descriptively convincing than dual federalism, it also carries strong, if less obvious, ideological and symbolic overtones. This theory meshes with pluralism and, by extension, with ideologies associated with freedom of choice and minority rights. Instead of the local, direct participatory access to government provided in the dual federalism perspective, the system has created a complex of institutions to accommodate the political reality of plural interests in the multicentered perspective.

In multicentered federalism, interests are inevitably plural and heterogeneous. Because no single government agency can respond to these diverse, conflicting interests, many governmental agencies arise instead. Together they constitute a jumble of governance that not only contains multiple goals but also accommodates their inevitable conflicts. If an interest group is dissatisfied with a dominant agency, it will find recourse in another agency, another level, or another branch of government. New programs, legislation, and governmental agencies develop in response to these demands.

This very complexity, rather than nostalgia for simple town meeting democracy, is widely regarded as an effective vehicle for citizen access because it provides "multiple cracks." Rather than presenting a single answer applicable to all, the complexity offers varying answers. A system with many points of access and many sources of remedy is sure to be more responsive, at least to minority concerns, than a simple unitary system.

This openness in multicentered federalism means that power is neither centralized nor fixed in a check-and-balance system of discrete autonomy, but rather is located in multiple agencies, groups, and persons and coalesces into varying coalitions around special problems (Dahl, 1961). Moreover, the fluctuating power relations, many constituent parts, and self-interested behavior fashion innumerable combinations and recombinations over time and space to create an overall, unplanned response to change and diversity.

Thus, multicentered federalism suits America's endemic complexity in motion. Processes of balancing, countering, bargaining, and splintering of interests are reflected in interactions among organizations. This interplay is analogous to a market of governmental agencies, generating new activities and agencies in response to diverse demands while balancing multiple interests in a collective, "steady state" equilibrium.

In short, multicentered federalism has ideological and symbolic connotations because of its compatibility with the dominant political theory of pluralism and, by extension, with the ideals of freedom of choice and minority rights. Like the dual federalism model, its ideological ties arise from localism. The multicentered theory takes its meaning not from the discrete autonomy of local government, however, but rather from the composite welter of fragmented local government.

Perhaps because functional federalism originates from a centralized rather than the particularly American localist perspective, it has few ideological proponents. Except for a few who see centralized federalism as beneficial in coalescing the country in pursuit of national purposes (Sundquist, 1969), such as equal opportunity, most of the writing on functional federalism either simply describes the phenomenon or views it with alarm. In calling for strengthening the states, both recent political rhetoric and old traditions have tried to reverse centralization to restore dual federalism (Kestnbaum, 1955; Lowi, 1969, pp. 305-306).

Much of the concern over functional federalism stems from its specialized professionalism (Bledstein, 1978). Beer (1977) considers the interaction between "confederal" (i.e., dual and multicentered) beliefs and functional federalism reality in this light:

> The confederal model offers the solution of dividing the bureaucratic forces and dispersing control over their incentives among several layers of government linked by contract. The reality, as the American experience with functional federalism amply attests, is that such a classic confederal tactic is weak and even dangerous in the face of the tendency of professionals to work together across intergovernmental boundaries for policy aims of their own. . . . While the confederal approach fragments the possibility of democratic control, the technocratic guilds, entrenched . . . will have their view of public interest prevail. (p. 33)

Specialization secures the powerful ties among the administrative agencies, legislative committees, and interest groups (Beer, 1976, pp. 166-171; Landau, 1969; Seidman, 1970, p. 37). Frequently, the interest groups are major industrial concerns.

Agribusiness, housing interests, highway interests, labor unions, medical professionals, and many other private interests are tied to administrative agencies by their respective specialized function, which tends to keep them fragmented.

The specialization of functional federalism ensures that many corporate-professional liaisons may pursue their special, highly differentiated interests without a concerted strategy among them. Because functional federalism poses decision making by multiple functions, each quite separate, its basis in specialization precludes a single locus of power.

This last point underscores the key feature of federalism: built-in protection against unitary rule. All the competing theories contain structural but different divisions of governmental power. Dual federalism divides by territorial level, and functional federalism divides by specialty. True, responsibilities are shared in multicentered federalism; nevertheless, the constantly changing combinations and decision points of this theory of fractionating federalism constitute a different but genuine sort of division by time. Thus, each prevailing theory presents a competing structure for separating power.

Despite their contradictions, all three theories are commonly held views of governmental relations. To sort out how the intergovernmental system works, the following chapters examine governments' interdependent activities, patterns of interaction, and basic structure and dynamics. In the process, discussion untangles contradictions among the three dominant theories of the intergovernmental system and shows some aspects of each theory that remain important.

NOTES

1. See Wright (1990) for further conceptual comparisons.
2. Wright (1988) makes a similar set of theories that he calls "coordinate," "overlapping," and "inclusive."
3. Grodzins (1966) is usually understood to be the founder of this school of thought. The originator of the marble cake metaphor, however, is McLean (1952), who coined the term.
4. Despite this integration, each vertical function is not so monopolistic as to be either closed or static. Consequently, the pickets of government multiply in parallel (Landau, 1973). For example, employment efforts by federal, state, and local government chains are approached through the capital-development channel (Economic Development Administration, Department of Commerce), the human welfare channel (Social and Rehabilitation Service, Department of Health and Human Services), and the labor channel (Employment and Training Administration, Department of Labor). The various chains of organizations and activities elaborate government.

An alternative interpretation is that each chain of federal, state, and local agencies linked by its special function is monopolistic because clients for its service have no options. But in the United States, governments and chains of governmental agencies are redundant. Even the most extreme example of monopolistic public institutions, prisons, offers a little variety: White-collar criminals may go to comfortable minimum-security prisons.

5. The report's perspective, though, adheres to the model of dual federalism.

CHAPTER 3

INTERACTIONS AMONG GOVERNMENTAL INSTITUTIONS

Chapter 2, "Competing Theories of the U.S. Intergovernmental System," set out three dominant theories of intergovernmental relations. This chapter concentrates on current intergovernmental behavior. It conveys an image of many overlapping interdependent units constituting an intergovernmental system. This chapter describes government agencies' subdivided work and interactions.

Brief comments on systems and interdependence introduce discussion on the interdependence of public organizations. Next, the chapter describes specific types of interactions among government units. Categorizing these units not only shows the range and variability of governmental interactions but also describes the basic processes that create the larger intergovernmental system.

INTERDEPENDENCE OF PUBLIC INSTITUTIONS

Understanding the kinds of interdependence that governments display helps one untangle theories of intergovernmental relations and build a new interpretation of governmental organization. U.S. governance may be understood as a system of interdependent parts. Seeing the processes and patterns of relations among those parts is crucial to understanding a system. This system viewpoint should be distinguished from a traditional viewpoint on organizations.

A traditional private organization tries to achieve its goal,[1] which is understood and discrete. Even when a private organization, such as a diversified multinational organization, has numerous subgoals, these contribute to a general goal such as profit. These goals are specific, narrow, and partial. Each individual private organization has a unifying goal, and many private organizations collectively fragment goals.

The problem of an institutional framework for government is entirely different from the typical organizational design problem because the system represents the entire polity. A single traditional organization can evade goal controversy because it is a part, whereas the intergovernmental system is tantamount to the whole.[2]

Although the intergovernmental system may contain organizations with identifiable, discrete, partial goals, the system encompasses them all and so lacks composite clarity. Its goals are neither discrete nor partial, but rather as broad as the polities choose. Furthermore, except for survival (which becomes salient only in such extremities as world war), no national goal is so overarching that it subsumes and orders subgoals.

Government agencies, which appear to have partial, discrete goals, usually take on some aspects of the whole. The goals of public organizations are "multiple, conflicting, and vague" (Wildavsky, lectures at the University of California, Berkeley, 1976). Although this situation results from political compromise, it also permits flexible response to varying interpretations of holistic government goals such as justice.

When a public organization attempts to convert its general mandate, such as "decent housing and a suitable living environment for every American," into operational goals, alternative formulations usually harbor conflicts between such overarching goals (Dror, 1968). Creating a suitable living environment out of an obsolete, crowded slum, for example, may mean demolishing some housing. Even on a more concrete, immediate level, an agency's multiple objectives tend to conflict. For example, a governmental agency devoted to efficient, effective road production must also attend to expensive labor laws, environmental protection, and equal employment opportunity (O'Toole, 1993).

Because the intergovernmental system serves and represents the whole polity, its goals are bound to be both contradictory and nebulous. Unlike a conventional private organization, the intergovernmental system cannot have the luxury of clear, discrete, partial, and hierarchically related goals.

In operation, the system not only contains multiple goals but also accommodates their inevitable conflicts through organizational complexity and specialization. The system accommodates conflicting goals by segmenting them according to special interests. Fragmented purposes may be as narrowly benefiting as industry-specific protective tariffs or as broadly benefiting as the food stamp program, which serves not only the complex agriculture, food-processing, and food-distribution interests but also a large, widely based clientele: the poor.

In sum, unlike a conventional private organization, the U.S. intergovernmental system cannot order its numerous agencies that have contradictory goals. Moreover, because individual governmental agencies are bound together in a system, they constitute each other's environments.[3]

These specialized agencies interact in response to critical uncertainties in their environment of public demands for services, public suppliers of resources, and "crises" (internally manipulated and externally imposed). Like conventional organizations, they adjust to their environment; but, because their environment consists largely of other public institutions, their mutual adjustment takes the form of intergovernmental exchange (see Figure 3.1).

For example, a community college district has multiple goals, including vague ones about community service and vaguely particular ones such as successfully terminating students in an employment retraining project. The district's relevant environment includes the mayor's office interest in showcasing its performance, its teacher employees, its client-supplier, the state office of employment development, the community college board, its funder the state department of education, its funder the Private Industry Council (in turn funded by the U.S. Department of Labor), other training and placement programs, the extension

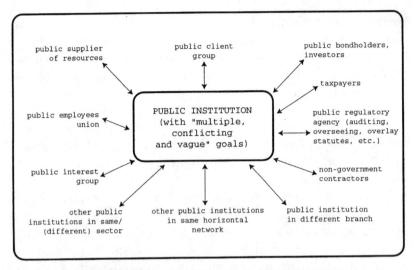

Figure 3.1. Environmental Demands and Uncertainties, Public Institution

school of the state university, and assorted interest groups such as Asian ethnic associations.

In operation, a public agency adjusts to the varying conflicting opportunities and demands posed by other organizations in the intergovernmental system. Depending on the issue, the governmental organizations involved are apt to share some of their multiple, conflicting, and vague goals with other governmental agencies in the particular problem constellation. Because the organizations share goals and because they impose continuing demands and constraints on each other, the boundaries between a public organization and its environment are diffuse. For example, some local redevelopment agencies once asked U.S. Housing and Urban Development (HUD) officials to interpret program guidelines as prohibiting a local agency from acceding to a neighborhood request. Within the larger intergovernmental system, the actions of one element shape the behavior of another in a continuing process of mutual adjustment.

To summarize, public institutions are interdependent and constitute each others' environments in the larger governmental system. Their mutual adjustment takes the form of intergovernmental exchange.

Before describing the various forms that mutual adjustment among governmental institutions may take, I should discuss a further aspect of interdependence: task division.

TASK DIVISION AMONG GOVERNMENTS

Task division structures key patterns of intergovernmental relations. If the tasks of government have become too complex for any one organization, how are the responsibilities for public work allocated among multiple governmental agencies?[4]

The authority for dividing governmental tasks, the constitution, gives scant guidance. Although it assigns states plenary powers over domestic matters, the constitution's 200-year-old design is too open to provide fine-honed categories for sorting out today's activities. No convention has developed to dictate which level may or must provide a good or service. Rules nevertheless abound:

- The lowest level is best; it is closest to the people.
- The highest level is best; it can best redistribute for equality.
- The highest level is best for some economies of scale.
- The lowest level is best for economic competition.
- The national government level should provide goods and services (e.g., welfare, health care, education) to people who are mobile; local government should provide place-grounded goods and services (e.g., water and sewer facilities).

If the constitution provides little guidance on task division and if no overarching principles have evolved to fill this breach, how does the intergovernmental system operate to allocate responsibilities among governments? The following discussion explores hierarchy, shared responsibility, and task technology as ways of allocating responsibilities.

Governments can be seen in a hierarchy proceeding from the base of cities to counties, states, and up to the apex of national government. This hierarchy may be steep, with each level successively delimiting the scope of the next level. For example, the national level could establish general rules; states could add restrictions or benefits; counties could manage welfare disbursements; and cities could help those who slip through the safety net. Or the hierarchy may be flat, with the national level providing more or less direction to states or cities. For example, the national government requires states to protect their environments; if they fail, the national government will act on their behalf. Regulations may check every imaginable contingency or fund abuse or offer wide discretion. Although some scholars see hierarchy in U.S. government, many do not, and few public officials would describe the U.S. government in terms of hierarchy.

Some government programs operate through shared responsibility. In a small, northern Wisconsin town, for example, the same people who wittily deride government as evil, inept, and lazy are rightly proud of the high-rise apartment house they built for some needy, snowbound senior citizens. No matter that funding came from the abhorrent national government. The town's volunteer leaders fashioned themselves into a proper housing authority, complied with countless requirements, and did good. The project was not the "government's," but the people's own.

When many different agencies with many different officials participate, it is easy to shift blame—nameless federal officials set requirements—and to take credit. Groundbreaking ceremonies for a project bring heads of agencies from every level, mayors, and members of Congress. All say they brought the project to the people, and in some respects they all did.

Sometimes the task technology may seem to dictate the task subdivision among government agencies. When the technology is well understood, the arrangement of the subdivided tasks logically flows from this technology. Relations among the subdivided tasks may take three forms (Thompson, 1967, pp. 51-65).

First, when its inputs, methods, and outputs are understood and predictable, then the task may be broken into detailed, specific instructions. This formalization permits decentralization without granting discretion to the multiple subunits. So, for example, the Social Security Administration can have many field offices while maintaining standards of fair, equal treatment. The outputs are pooled to achieve the overall objective. The field offices contribute to the whole, a nationwide program of equitable support for the elderly, but the field offices do not contribute to each other's work.

Second, subdivision arrangements may sequence tasks to achieve the overall objective. The initial task contributes to the second. To illustrate sequenced tasks: Land is plotted, roads and water mains are laid, and finally housing is built.

Third, subdivided tasks may be reciprocally interdependent so that the actions of each are shaped by the other. In already developed areas, for example, increasing intensity of land use often calls for increasing transportation capacity (both highways and mass transit), which in turn permits more intense land use.

When a task is not routine, it cannot be controlled through standard rules and procedures. A nonroutine task may be either novel, therefore not yet understood, or so dependent on varying circumstances in its environment that it cannot be systematized. Nonroutine tasks must be delegated more or less "whole" because the delegating agency has no knowledge about how to subdivide the task (Perrow, 1970). Then experts can develop the necessary understanding. So, for example, many research grants are given to generate workable ideas on alternative sources of energy. Local officials can adapt an activity to their particular environmental conditions. Thus, community development decisions, such as which neighborhood to rehabilitate and how, are left to cities to tailor choices to their unique physical and political circumstances.

Table 3.1 illustrates the division of public tasks with routine and nonroutine technologies.

The way the overall objective is formulated may substantially change the task technology and the logical arrangement of the task. The objective may be cast in a routine framework, or it may be presented as an innovation demanding new or fluid organizational arrangements.

For example, the problem of poverty may be formulated as "poor, elderly people have insufficient money." Accordingly, the solution is to increase their money. The solution is cast into a routine of writing checks according to standard rules, operating through Social Security Administration field offices. Rare circumstances or crises requiring deviation from the standard check are appealed upward through the hierarchy, and changes in check writing standards are directed downward to the field.

Alternatively, the problem of poverty may be formulated as "poor people have insufficient power." Accordingly, the solution is to increase their power. But such a solution cannot easily be cast into a routine. Although many plausible ideas may be proposed, no straightforward method for increasing power has been tested and proved effective. This situation, where the solution technology is both unknown and obviously dependent on different power contexts, defies routine.

Instead, it demands nonroutine innovative activities to respond to different situations. Building the poor's power may mean job vans in Boston, boycotts in Atlanta, and consumer cooperatives in Boise.

TABLE 3.1 Illustrative Examples of Intergovernmental Arrangements Resulting from Task Subdivision, Varying According to Task Technology and Problem Formulation

Task Technology	Problem/Solution Formulation	Organization Variables	Examples of Institutional Arrangements
Known technology and consequences	Routine, negative (regulation)	Flat hierarchy (wide span of control), high formalization for standardization	Federal-state point-source air pollution control; local housing codes
	Routine, positive		
	Pooled	Decentralized (discretion within standards)	Federal-state-local welfare checks
	Sequenced	Centralized planning, scheduling; discretion on task process, not output	Local development (e.g., roads, water, sewage, utilities, housing)
	Reciprocal	Central authority only to resolve disputes; lateral coordination through mutual adjustment	Transit—land use
Technology and consequences contingent on differentiated environments	Nonroutine, responsive	Decentralized decision making	Community Development Block Grants
Technology and consequences unknown	Innovation	Centralized resource allocation; decentralized decision making on experiments	Nonpoint source pollution

Criteria of innovation and responsiveness, coupled with a political bias toward grassroots participation, suggest a decentralized, "string-free," experimental organizational arrangement. That was precisely what was designed and formalized into one of Lyndon Johnson's first war on poverty programs: the Community Action Program.[5] Such organizational forms work through antiroutine, antibureaucratic, flexible decision rules. Political power, responsiveness, bargaining, and consensus building determine decisions.

These examples illustrate how different decision rules and task arrangements derive from variable formulations of the problem. The way a problem is posed depends on historical circumstances and political intent. As the problem of poverty illustrates, the same problem may be articulated in different ways at different times for different people. The lesson is twofold. First, division of government tasks is not preordained and may vary. Second, the way particular tasks are subdivided flows neither from the task nor from organization theory precept. Instead, it derives from the haphazard amalgam of historical circumstances, program evolution, court interpretations, and political expediency.

Over the years, new governmental responsibilities, programs, and services have been added to and complicated old forms of task division. Activities that might once have been the exclusive responsibility of a single unit have become shared. For example, local school districts once solely determined primary education in the United States. Now, states have major decision-making roles and in some places set funding levels, whereas the national government supplies crucial funding and corresponding requirements. Nonetheless, education is still popularly viewed as a local responsibility.

Sets of interdependent tasks frequently have not been planned but rather have emerged over time. Depression-era national housing programs, for example, had to mesh with various preexisting local housing codes and zoning. Furthermore, when an external crisis makes an issue politically salient, new tasks are apt to be layered onto an older set of subdivided activities. In the mid-1970s, for instance, an oil embargo and the formation of the Organization of Petroleum Export Countries (OPEC) generated new tasks (such as stimulating development of alternative sources of energy) layered onto the old tasks divided among public utilities commissions, the U.S. Department of the Interior, and other agencies. Soon after, the associated activities were programmed, transformed, and made prominent into a new energy sector.

In sum, the complexity of modern governance induces division of tasks among governmental agencies in a range of ways, depending as much on the politically formulated problem as on the technical nature of the task. Particular intergovernmental arrangements tend to derive more from historical circumstance and short-term political cross currents than from conscious design. Taken together, the various task subdivisions and shared, overlapping goals compel interdependence among governmental agencies. Their mutual adjustment takes the form of intergovernmental exchange.

INTERACTIONS AMONG GOVERNMENTAL AGENCIES

Exchanges between governmental agencies may be either collaborative or conflictual. They may take various forms, ranging from autonomy to total merger on the collaborative spectrum, and from autonomy to counter programs and funding on the conflictual spectrum. The following describes prototype interactions between government agencies.

Collaboration Among Governments

Figure 3.2 shows a spectrum of collaborative forms of mutual adjustments[6] between two hypothetical governmental units. The following spectrum sets out nine kinds of collaborative arrangements, which are then briefly described and illustrated. Intergovernmental collaboration characterizes much of government work and is widely and well documented (e.g., Chisholm, 1989) and considered routine (Agranoff & Rinkle, 1986). Moreover, some argue that intergovernmental arrangements can be matched to policies to achieve the most effective collaboration. (See O'Toole, 1993, for assumptions and reservations about this approach.)

Autonomy

This extreme form means governmental institutions have no effect on each other whatsoever. One example is the mutual autonomy of the U.S. Navy and Kansas City's Regional Railroad Commission, which have little need for collaboration. A second example suggests that officials may think of their agencies as autonomous, when in operation they are blindly collaborating: A local community college gives classes to displaced workers whose English is poor. The college receives funds from the state department of education according to a formula for average daily attendance in class. At the same time, the college receives funds from the U.S. Department of Labor for retraining the unem-

Interactions Among Governmental Institutions 33

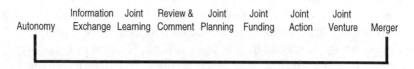

Figure 3.2. Spectrum of Intergovernmental Collaborative Arrangements

ployed. Yet, presumably neither the national Department of Labor nor the state department of education is aware of the collaboration.

Information Exchange

This minimal form of interaction is the simple exchange of information. The information may be facts about an agency's interests and activities or data on its clients, services, or products. Agencies exchange information to promote their own efforts, to learn new approaches, and to identify potential problems and opportunities. Information exchange implies only that the organizations have the opportunity to adjust to each other; it does not compel them to do so. Agencies may screen out information they receive from other agencies without absorbing it, or they may note it but not take it into account when undertaking their own activities. Examples of information exchange include the County Supervisors Association of California newsletter and cocktail parties after government conferences.

Joint Learning

Joint learning refers to the process whereby governmental institutions discover together something new that neither knew or realized before. Joint learning may be generated by pooling partial information each institution already had or by a third, external party's either providing the new skill or knowledge or structuring the learning experience. Joint learning means the new understanding is shared by the governmental institutions; the fresh ideas originated in and influenced them both.

What is learned can range from minor facts or techniques to major trends and theories; thus, joint learning may be joint problem defining. Examples of joint learning include conferences with active workshops and hearings with intergovernmental deliberation prior to policy enactment. Through such a vibrant exchange, for example, technical and political officials together learned an entirely new conception of rent control: Analysts pooled their economic interpretation of rental rights with politicians' interpretation of rental rights drawn from constituents who had lived in the same building for more than 50 years.

In another example, the San Francisco Estuary Project (under the auspices of the federal Clean Water Act) brought together representatives of federal, state, and local governments as well as representatives of key private interests, such as environmental and business. Despite considerable differences of interests, after the investment of much time and social, intellectual, and political capital and through the services of skilled professional facilitators, the group engaged in joint learning. Members reached agreement on water quality indicators (e.g., salinity level) for the estuary (Innes, Gruber, Neuman, & Thompson, 1994).

Review and Comment

Review and comment means more interaction between governments because one organization may influence not merely the understanding but also the action of another. These procedures require the initiating agency to clear a proposed action with other governmental units that could be affected by that action. Although many review and comment mechanisms carry no veto power, they customarily require the initiating agency to respond to comments and in essence to justify an action that defies the preferences of other governments. In this way, review and comment interactions trade off the interests and values of government agencies. Examples of review and comment include the old Office of Management and Budget circulars A-95 (intergovernmental review of applications for federal aid) and A-85 (intergovernmental review of proposed federal government actions), the *Federal Register* (comments on regulations), and program-specific certification procedures. Six states have established intergovernmental review procedures for major projects, known as *developments of regional impacts* (Morris, 1997). A city could submit its industrial development plan to intergovernmental review and then modify the plan's transportation routes to respond to comments from a regional parks commission.

Perhaps the most familiar review and comment process is environmental impact statements. This procedure invites comments on potential environmental effects of proposed projects and proposes mitigation measures. The process is valuable in increasing the influence of internal analysts (out of fear of outside comments) and providing a political forum for objections. The process's wide range of participants and high volume of criticisms tend to preclude gentlemen's agreements (Taylor, 1984, pp. 183-184).

Joint Planning

Joint planning means sharing the process of setting out a subsequent course of action. More than one government participates in developing proposals. By planning together, governments interact and impinge on each other to a greater degree than in the more detached review and comment arrangement, but they may continue to act separately after their shared planning. They may jointly devise a plan that can be carried out by only one governmental organization or by several acting independently. So, for example, the state Coastal Conservancy, the metropolitan Bay Conservation and Development Commission, and the city Office of Parks and Recreation jointly planned for Berkeley's waterfront's preservation and use. All the governmental agencies had jurisdiction over and interest in the waterfront's condition. Other examples of joint planning include joint resources management such as river basin planning, coastal zone planning, and the Natural Communities Conservation Planning Program.

Box 3.1

CALFED offers an example of joint planning. It consists of five California and five federal agencies with management and regulatory responsibilities in the San Francisco Bay and the Sacramento Delta. The joint planning also involves stakeholders such as urban and agricultural water users, fishing interests, businesses, and environmentalists. The planning addresses such issues as ecosystem quality, water quality, water supply reliability, and Bay-Delta system vulnerability (CALFED, Bay-Delta Program, 1996).

Joint Funding

Joint funding means a commitment of resources by several governmental agencies. These organizations are mutually bound to and accountable for the activity and so more interconnected and "interinvested" in the collaboration than in a joint planning arrangement. Actions may be performed separately, undertaken by only one organization but jointly supervised, or delegated entirely to one agency.

The organizations may fund separate tasks and so isolate their accountability, or they may pool their resources to be reallocated among all the tasks. This allocation depends on the formal purposes each contributor may legally fund. Some agencies may have broadly stated goals or discretionary funds, and thus more flexibility than agencies with more narrowly construed objectives. To protect fiscal accountability, jointly funded activities invariably entail documents (ranging from fairly informal interagency memoranda of agreement to full-scale contracts and formula-based legislation) on the contribution of resources and the purposes and projects of the activity.

Examples of joint funding include national grants matched by nonnational shares, integrated grants by several national agencies, and local government dues to regional councils of government. For a specific example, a local agency may receive community development block grant funds to rehabilitate dilapidated housing and Job Placement Training Act funds to teach teenagers carpentry by practicing on housing repair.

The North American Water Fowl Management Act (from a treaty between the United States and Canada in 1986 and the United States and Mexico in 1994) offers interesting examples of joint funding. As part of a general effort to recover and preserve waterfowl, U.S. Fish and Wildlife facilitates the establishment of working groups in the various waterfowl flyways. In the Central Valley Flyway, a number of independent agencies such as the Environmental Protection Agency, California Fish and Game, and nongovernmental organizations such as the Nature Conservancy and Ducks Unlimited coordinate their efforts. Their activities include joint funding for acquiring properties for conservation and restoration (North American Waterfowl Management Plan, 1995).

Similarly, a number of government agencies can jointly receive funding. For example, Alameda County, California, facilitated a number of cities and a variety of nonprofit service providers to prepare a joint application for McKinney Act funding from HUD. It received the largest supported housing for the homeless grant nationwide and managed the grant activities though a coordinating committee.

Joint Action

Joint action means sharing not merely planning and funding but also the conduct of an activity itself. Typically, governmental units work together by subdividing operational tasks. Joint action through subdivision of tasks among governmental units may take a wide variety of forms depending on the nature of the problem, the task technology, and preexisting institutional arrangements and capacities. For example, national, state, and regional agencies divide tasks to improve the environment. The result of these often complex divisions of labor is completed public work.

Joint Venture

More rarely, officials from different agencies work hand in hand. Federal Bureau of Investigation officials may work with local police on a special case. In natural disasters, officials from the U.S. Department of Health and Human Services help Federal Emergency Management Administration officials, local firefighters, and public health practitioners. In such joint ventures, tasks are not subdivided and allocated to separate governmental units but are shared. The shared resources include not just funds but actual powers and personnel of the participating governmental units.

Forms of joint venture may be distinguished by degree of investment and permanence. A joint venture may be a short-term, part-time, ad hoc intergovernmental task force engaged in resolving a specific problem, for example, work fare standards. Alternatively, it may be a longer-term, full-time intergovernmental team addressing multiple aspects and ramifications of a new development, for example, offshore drilling. Or a joint venture may be a quasi-permanent group developing and administering new joint activities. For example, the Lower Great Lakes-St. Lawrence Basin joint venture involves U.S. funding from the North American Wetlands Conservation Act and from the states of Vermont and New York to protect 6,000 acres of wetlands and adjacent uplands (North American Water Fowl Management Plan, 1995).

Participants in joint ventures bring the powers and resources of their home governmental units to the collective effort, and so are bound by and accountable to parent institutions. At the same time, they are collectively responsible for the joint venture. This dual responsibility places some tensions on participants, who must balance the objectives and perspectives of parent institutions with those of the immediate collaborative effort, often a crisis.

A less extreme version of joint ventures occurs when members of one governmental institution serve on the boards of others. They bring the perspectives and often the decisions and resources of their parent institutions to the host, which in turn adjusts its activities to accommodate the guests.

Regardless of the method, degree of investment, purpose, and duration of the joint venture, it integrates the participating governments through their collective participation in the particular effort. The contributions of powers, resources, and personnel to the joint venture tend to affect the ongoing internal operations of the parent governmental institutions. In the Coachella Habitat Conservation Plan, the U.S. Fish and Wildlife Service, the Bureau of Land Management, and California's Department of Fish and Game all contributed funds or land for habitat acquisition and share management.

One broad-based example of a joint venture is undertaking a neighborhood facility, that is, designing, building, and operating a multipurpose community center in a poor neighborhood. With HUD community development block grant funds, local government deals with the building and service organizations, handling diverse activities such as drug abuse, job placement, day care, bilingual education, and elderly nutrition. Each service organization is surrounded by a network of federal, state, and local agencies concerned with the funding, regulation, and operation of its service. Each governmental agency contributes to the center's eventual operation. The joint venture depends on collective interaction.

Merger

This extreme intergovernmental collaboration means that the institutions merge their purposes, powers, resources, and personnel for all activities. The merger constitutes a new, distinct organization with its own guiding mission and authority. Merger means that the participating governmental units lose autonomy. The very term *intergovernmental* loses its meaning because nothing can occur between integrated units.

Nonetheless, even a merger is relative. Functions may remain with the components, which might thereby retain clusters of personnel with their distinguishing history and operational style. These components might even preserve some remnants of their former identity. For example, in HUD, formed through the merger of several agencies, the assistant secretary for housing retains a second title from the prior era of autonomy: commissioner of the Federal Housing Administration (FHA). The old FHA legislation still exists; its work is ongoing and its lore continues its tradition.

The Exxon-*Valdez* oil spill generated a particularly interesting form of intergovernmental merger. Part of the settlement of the civil suit involved $900 million to restore the tidal marshlands. Federal and state governments fought over which government was responsible. Because both had the authority and obligation, eventually they decided to be cotrustees (three federal, three state) and share responsibility for dealing with the oil effects. They are acquiring land (to reduce other effects on the tidal marshlands) jointly. For example, the U.S. Forest Service buys some land, but with a state easement. Their shared legal ownership of the property acts as a check on each other (General Accounting Office, 1993).

Conflict Among Governments

Current events, long-term trends, political pressures, and structural competition may make one governmental unit oppose the actions of another. These conflictual reactions mirror collaborative interactions. A range of divisive mechanisms counters the range of collaborative arrangements. The divisive mechanisms are not merely the absence of collaborative arrangements but actions that assert the separateness of the governmental organizations and emphasize their differences.

Figure 3.3 shows a spectrum of increasingly divisive forms of interactions between governments with six forms of intergovernmental obstruction.

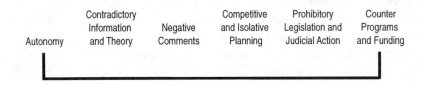

Figure 3.3. Spectrum of Intergovernmental Obstructions

Autonomy

The mildest form of intergovernmental obstruction, autonomy, means merely ignoring the actions of other units, either considering them irrelevant or trivial or trusting that the objectionable actions will be temporary.

Governmental units protect their autonomy by buffering themselves with high revenues. Thus, some suburbs can afford to refuse national school aid to avoid Department of Education equal opportunity requirements. Others buffer themselves from other governments by defining their unique, elite expertise in a way that cannot be challenged. Until recently, for example, the Central Intelligence Agency and the Atomic Energy Commission preserved autonomy through specialized knowledge. These subtle forms of intergovernmental obstruction can be quite powerful. The most impressive case in point is the Department of Defense, a virtual monopoly that asserts and protects its privilege through both resources and expertise. Yet, no governmental unit is entirely autonomous. Today, the Department of Defense is deeply involved in intergovernmental negotiation and planning with local governments about defense base conversion to nonmilitary uses.

A public university attempts to buffer itself from political pressure to protect its integrity and academic freedom not only through constitutional protections but also through funding. It reduces its dependence on state funding by relying on extensive research grants (from diverse federal agencies) and developing a substantial institutional endowment from alumni donations and foundations.

Some theorists, such as Wilson (1989), argue that agencies prefer autonomy over competition for turf expansion, bigger budgets, and responsibilities because they are liable to increase complexity and constraints. Kunioka and Rothenberg (1993) use the U.S. Forest Service and the National Park Service as an example to show that these near functional rivals avoid competition. Their Mono Lake case shows agencies passively avoiding responsibility for a tricky environmental and political situation.

Contradictory Information and Theory

This form of intergovernmental obstruction is the opposite of information exchange and joint learning. By not only withholding information but also publicizing information and theories that counter those put forth by another organization, a governmental agency may obstruct the actions it opposes. For example, the Federal Reserve Board may issue interpretations of economic data that counter those of the Office of Management and Budget. Information becomes a strategic weapon.

The Exxon-*Valdez* Oil Spill Council illustrates not only merger but also competition through 10 years of contradictory information and theory. Both the federal and the state governments wanted control over the $900 million turf. Some theorists (Downs, 1966; Niskamin, 1979) argue that interagency competition for resources and responsibilities is the norm. Bardach and Lesser (1996), for example, find turf barriers to interagency collaboration in human services delivery.

Negative Comments

One governmental agency's use of review and comment procedures (e.g., environmental impact statements) to obstruct another's action is common. Even though review and comment systems were designed to expand information, they can easily be used to block undertakings. For example, if one city proposes to build a shopping center, adjacent communities may object on "legitimate" bases, such as undue traffic spillovers or violation of the regional growth plan, but their intention may be to sabotage the project to keep business and tax receipts up in their own downtown business districts.

Competitive and Isolative Planning

This way of obstructing actions of other governments uses project and long-term planning competitively. Federal categorical grant-in-aid programs take advantage of competition to generate quality applications.

With limited funds available, only the best proposals win grants, and so the program's chances for success are relatively high. Despite an increasing reliance on entitlement programs (which give funds to city and state agencies as a right, rather than something earned) and the current vogue for devolution, competitive planning persists. Governmental agencies delay review and comment notifications and environmental impact statements as long as possible to reduce the threat of adverse comments. Similarly, parochial land use planning emphasizes a city's self-interest at the expense of neighboring cities. Using exclusionary zoning, such as large lots, to shunt undesirable poor people into other cities is a common practice for evading regional responsibilities.

Competitive planning for transit in Minneapolis between the Metro Council and the Metropolitan Transit Commission generated alternatives, enormous media coverage and public attention, substantial council members' decision making time, and delays. It also facilitated the maturation of a new idea of transit, which may be worthwhile (Bendor, 1985, pp. 161-166).

Prohibitory Legislation and Judicial Action

Governmental agencies resort to legislatures and courts to obstruct another agency's actions. This extreme approach may be limited to a single case or an entire class of governments. In the case of *Milliken v. Bradley* (1974), a suburb argued it need not integrate its schools. For another example, California enacted a law that requires favorable local government referenda before public housing can be built. Local citizens must actively vote for public housing, rather than letting officials make the decision, as is customary in other states. This obstructive requirement was thought to slow public housing production. Negative legislation is not limited to local governments resisting higher-level impositions. For example, at one time when the city of San Francisco voted to prohibit guns, the state legislature passed legislation preventing such local ordinances. (California cities now may—and some do—prohibit certain types of guns.)

Lawsuits may emerge from a single intergovernmental conflict or alter the entire national intergovernmental mode of carrying out a program. For example, the city of Emeryville sued California's Department of Transportation for overlooking it in an environmental impact statement for a freeway. Emeryville received a freeway off-ramp to its retail park from the settlement. In another case, a suit against the Environmental Protection Agency transformed implementation of transportation controls nationwide (Howitt, 1984).

Counter Programs and Funding

When the action of one governmental institution is sufficiently objectionable and seemingly intractable, it may be countered by affirmative programs and funding for the opposite effect. The national government alone offers many examples. Franklin D. Roosevelt established an entire bureaucracy of "alphabet" agencies to counter the existing Republican-dominated organizations. More recently, the legislature in a southern state passed a provision supporting private schools to counter national efforts at school integration.

Because the intergovernmental system is not hierarchical, conflicts cannot be simply passed up to a higher level for a ruling. Reflecting its true complexity, the intergovernmental system instead contains many interactive mechanisms for reconciling disputes. Important ones are associations of local governments in metropolitan areas (e.g., which might help adjoining communities deal with their shared lagoon) and ad hoc intergovernmental task forces around specific problems such as land development and watershed preservation (Bollens, 1992). Innes et al. (1994) argue for stronger formal structures for consensus building. However a particular dispute is managed, it occurs in a larger context of ongoing intergovernmental relations. Thus, the same agencies may conflict on some issues and collaborate on others.

This range of intergovernmental obstructions suggests that governmental interaction may be divisive as well as collaborative. Interaction itself influences the governmental organizations, preventing them from remaining static. In effect, the governmental interaction is a process of adaptation.

SUMMARY AND CONCLUSIONS

In sum, the U.S. government is a system of interdependent governmental agencies. These contribute to each others' environment—of critical resources, such as funding; constraints, such as regulations; and opportunities, such as new constituencies. Mutual adjustment between governmental agencies takes the form of intergovernmental exchange. Much of this interaction occurs over shared work and shared, overlapping goals. Interactions between agencies may take many forms, both collaborative and divisive.

This chapter has begun to describe the U.S. intergovernmental system as it operates after more than 200 years of development. Today's wealth of complexity seems chaotic. Each of the myriad governmental agencies has a number of vague, incompatible goals. In composite, they are overlapping, contradictory, and confounding.

This complexity correlates with specialization. Because government's work and political constituencies have become too complicated for any one agency, activities are subdivided into tasks that can then be tackled. But specialization also calls for ways to bring those parts together into some coherent achievement. Rather than isolating agencies, task division breeds their interdependence. Because agencies' goals cannot be rank ordered, and because their tasks are intertwined, governmental agencies are almost constantly interacting. This complex array of interactions is not random, but rather forms patterns. The next chapter introduces key dimensions for examining the intergovernmental system's structure and processes that shape those patterns.

NOTES

1. The goal may shift to justify and maintain the organization (Barnard, 1958; Thompson, 1967). W. Richard Scott (1992) outlines dominant theoretical models of organization theory that clarify competing views of goals and organization maintenance. Thus, natural systems theorists view the organization "more as an end in itself than as a means" (p. 5), whereas rational systems theorists view the organization as a means to achieve its formal goal. "Similarly, [w]hile a rational system view would insist that these two criteria [goal attainment and self-maintenance] are mutually supportive, natural system analysts note how easily and often they become independent if not inconsistent" (p. 11). (See Chapter 3, Note 2.) For further discussion and comparison of rational versus natural, closed versus open organizations, and an excellent bibliography, see Scott. The perspective of this book borrows a bit from both rational and natural models, but it is emphatically in the tradition of open systems.

2. Of course, the intergovernmental system operates in a larger system and interacts with other larger-scale elements in the environment, such as the structure of society. Just as a system perspective invites exploration of still-larger systems, so it invites exploration of smaller systems operating in its constituent elements. Thus, a private organization harbors its own divisions and interdependencies.

3. Organization theory (Biller, lectures at University of California, Berkeley, 1976; Lawrence & Lorsch, 1967; March & Olsen, 1976; Pfeffer & Salancik, 1978; Powell, 1990; Powell & DiMaggio, 1991; Thompson, 1967) has concentrated on new institutionalism, network theory, management of critical uncertainties, survival in an unstable environment, and other issues related to organizational dynamics subject to external conditions. The theory of the intergovernmental system developed in this book builds on those principles of contingency management, new institutionalism, and network theory but considers uncertainties and stresses, although external to the individual agencies, as factors generated within the larger intergovernmental system.

4. The processes of distributing various governmental tasks are discussed in Chapters 4, "Dimensions Structuring the Intergovernmental System," and 5, "Sectoral Dynamics: Institutionalized Technological Elaboration and Effects on Area Interests." They are complex and political (see Moe, 1989).

5. The Community Action Program was converted into community services and shifted to a less political and innovative orientation.

6. The discussion on collaborative arrangements extends and adapts a lecture by Pfeffer (lecture at the University of California, Berkeley, November 21, 1977) in which the joint venture and merger arrangements were discussed as a means for coping with critical uncertainties for business. See also Pfeffer (1972), pp. 382-394.

CHAPTER 4

DIMENSIONS STRUCTURING THE INTERGOVERNMENTAL SYSTEM

The last two chapters present a sharp contrast. Chapter 2, "Competing Theories of the U.S. Intergovernmental System" poses three competing theories of how U.S. governments do and should relate. Chapter 3, "Interactions Among Governmental Institutions," shows a wide range of intergovernmental relations in current practice. Although that jumble of governmental interactions somewhat resembles multicentered federalism, the image of intricate interactions forming a system seems more complex—and robust—than any prevailing theory. This chapter sets forth key dimensions for seeing patterns of intergovernmental relations. In doing so, it helps link theory and practice.

Organizing principles from the theories suggest analytic dimensions: vertical (the pickets linking federal, state, and local levels in functional federalism), horizontal (the territorial division of layer cake, dual federalism), and time (how the relations change over time, a factor implicit in multicentered, swirling, marble cake federalism). A fourth dimension, decision-making mode, helps analysis even though it is not inherent in any of the theories and may not be strictly necessary to the argument. Each dimension is examined to increase understanding of current governmental operations and to establish the framework for subsequent analysis.

THE VERTICAL DIMENSION

According to functional federalism, the U.S. intergovernmental system organizes by specialty, linking federal, state, areawide, and local levels in a vertical chain (Figure 4.1).

A common policy concern ensures interdependence among the units composing a vertical chain. The agencies are bound together by their shared problem, discipline, and professional perspective. Furthermore, they are bound together by their interlocking procedures. Some of these procedures and coordinating tasks are mandated by laws or are conditions for receiving grants (Mandell, 1990). Governmental agencies within each specialization rely on each other to complete their tasks. In relatively elaborated functional areas such as housing, interdependence between national actions (e.g., interest rates, National Mortgage Association decisions, and subsidies and tax credits for low-income housing) and local actions (e.g., bonds, zoning, and building codes) is crucial. Even simpler functional areas such as law enforcement display some interdependence. Although localities depend on the national government for funds and special research, the national government depends on localities to spend the money and reduce crime rates. Occasionally, the Federal Bureau of Investigation collaborates with county sheriffs and city police departments.

The vertical chains manifest some centralization and implicit hierarchical relations, varying in style and degree by policy area. In some instances, local institutions are no more than service stations (Derthick, 1970) or administrative districts of the national governments. Kettl (1988) writes of related and often complex arrangements as "government by proxy." In other arrangements—for example, water quality—states have some authority within constraints with the national government in reserve, ready to act should a state fail. For example,

Dimensions of the Intergovernmental System

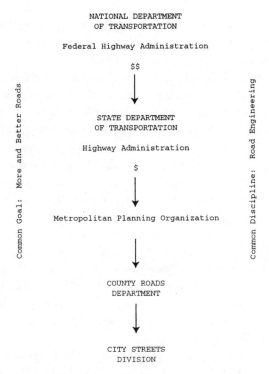

Figure 4.1. An Illustration of a Chain of Agencies Aligned Vertically

joint federal-state planning arose because California failed to plan for the San Francisco Bay-Delta water quality.

In still other policy areas—for example, community development—cities have wide decision-making scope, seemingly autonomous. Indeed, cities refer to the community development block grant as their own, even though it comes from the U.S. Department of Housing and Urban Development (HUD) with conditions on the use of that money. In still other policy areas, notably education, cities may be quite constrained by state and federal court decisions, state regulations and reporting requirements, and national funding requirements, and yet act independently. In some policy areas, all levels are involved; in others, only two or three play major roles. The extent of hierarchy and the degree of national direction in the various policy concerns follow no logical or overall pattern[1] but rather seem to take their form by historical accident and political happenstance.

Despite some evidence of hierarchy, relations among governmental units within their respective vertical chains scarcely resemble the superior-subordinate positions of a command structure. Instead, relations tend to be reciprocal. For example, officials in the national HUD, a state department of housing and community development, and a mayor's office of community development do not give and receive orders up and down the chain, but instead jockey ideas and opinions in all directions. Interactive, joint decision making is especially noticeable on issues of problem definition and technological change. A new finding, a successful innovation, or a disaster in one unit often triggers a reassessment of technology and so a readjustment of the relations among the units within the vertical chain.

The remainder of this book refers to the vertical chains of agencies, linked together by their common goal and reciprocal tasks, and buttressed by the legislative committees and private interests in their specialties, as *sectors*. Webster-Merriam's (1956) third definition of a sector, "a subdivision of . . . a system . . . as an area of responsibility" (p. 765), comes closest to this intended meaning, if the system is understood as the U.S. political economy.

Clearly sectors—multiple organizations linked together in the same policy area—are a strong trend (Kettl, 1988; O'Toole, 1993). From all the terms in current use,[2] (e.g., *networks, issue networks, interorganizational policy systems*) the term *sector* is chosen because it is short and it connotes ties with the new institutionalism (Powell & DiMaggio, 1991) and the larger political economy.

Existing sectors constitute established, institutionalized[3] responses to past political and economic demands. New sectors, energy, for example, emerge as new problems emerge (Lowi, 1969; McConnell, 1966). Figure 4.2 conveys sectors' multiplicity, evolution, and pattern of vertical alignment by specialty.

The policy sectors are widely recognized in practice and in the government and sociology literature. The distinctions dividing the different sectors are taken for granted (see Note 3). Mandell (1990, pp. 34-35) lists studies of interorganizational networks (sectors) in health and human services, human resources training, highway construction, and emergency management. Others include agriculture, community development, economic development, education, and housing.

This preliminary description of sectors provides a base for later exploring their dynamics and power. In the meantime, it shows how the

Dimensions of the Intergovernmental System

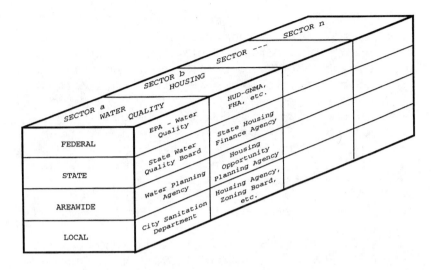

Figure 4.2. The Vertical Dimensions—Sectors

vertical dimension is crucial to analysis of the U.S. intergovernmental system.

Box 4.1
A Day in the Life of a Planner in the Housing Sector

A planner in a city's department of housing and community development (hereafter, the department) begins her day reviewing a student intern's work on the consolidated plan for housing required by HUD. She focuses on the section documenting housing needs. She phones the housing planner at the metropolitan council of governments, asking about possible changes in its calculations of her city's "fair share" of the region's need for affordable housing.

Next, she attends a staff meeting, which covers a new set of regulations from the California Housing Finance Agency and local nonprofit housing development corporations' problems with the department's revolving loan fund. Next, she goes on a site visit with the executive director of a local housing development corporation to a housing project stalled during construction. The department has contributed some of its HOME funds (a block grant from HUD) and is quite concerned that the project be completed.

Back at the office eating lunch at her desk, the planner skims the *Housing and Development Reporter* for a congressional update. She also glances at *Housing*, a journal for government housing specialists, putting aside the more academic *Housing Policy Debate* (published by the national quasi-independent Fannie Mae Foundation) for her subway ride home. After lunch, she completes a questionnaire for the National Association of Housing and Redevelopment Officials on how she finds and leverages multiple funding sources for housing projects.

At 2:00, she goes to the regional HUD office for a workshop on new McKinney Act (housing for homeless) regulations. During the coffee break, she meets with several of her graduate school classmates. After catching up on personal news, they discuss funding problems and how the new administrative people in the state office of housing and community development are getting along with the state legislature's housing committee. After the workshop ends, she takes the subway home while reviewing the workshop materials. (She forgot the *Housing Policy Debate* at the office, but it will keep.)

THE HORIZONTAL DIMENSION

Popular and traditional conceptions of U.S. government are rooted in territory. Moreover, territorial divisions are at the heart of the theory of dual federalism, and maps make these divisions deceptively obvious. The horizontal dimension for analyzing the governmental system is both easily defined and intuitively acceptable. Yet, because issues of intergovernmental relations concern interactions of governments, the horizontal dimension needs more description.

Governmental organizations cluster on a horizontal axis, that is, on the same geographical area, in two ways. They may have the same formal jurisdiction or not. The first, when the agencies serve the same political jurisdiction, may be easier to grasp. General purpose executive branch governments are composed of multiple special purpose departments and agencies. In addition, some autonomous agencies and commissions, such as housing authorities and the Federal Reserve Board, also have specialized executive responsibilities for the same geographical areas as their respective general purpose governments. At the same time, courts and legislative bodies serve the same geographical areas as executive institutions. Although less specialized than the executive branch, they too have subunits with varying degrees of autonomy.

Multiple governmental organizations serving the same jurisdiction are not related hierarchically. Therefore, they are incompatible with conventional public administration precepts for rational, efficient order. Multiple agencies cannot be centrally controlled because politics defies application of management techniques designed for private organizations. Even within the executive branch, a mayor cannot order city hall's departments to be subsumed and ranked under a single, overarching municipal goal, such as efficiency, because operationally the multiple departments' goals conflict and their constituencies differ.

As other, more autonomous authorities pursue their objectives without regard to the mayor's priorities, prospects for management diminish and rivalries increase. Conflict also arises between branches. Partisan politics encourages members of the city council to counter the mayoral agenda with their own reforms; judges may overturn entire policies and programs. These often competitive nonhierarchical relations illustrated at the city scale also occur at other levels, for example, between the president and Congress. Generally, even when governmental organizations serve the same jurisdiction, their different goals, constituencies, and perspectives on the public interest generate a range of complicated, unconcerted interactions.

The second, more complicated, kind of horizontal clustering occurs when multiple governmental institutions serve the same geographical area but do not share the same formal jurisdiction. Such intergovernmental collectivities are most easily perceived and frequently studied at the metropolitan scale. The typical "real city" metropolis comprises many general purpose county, city, and township governments. Even the moderate-sized Minneapolis-St. Paul metropolitan area has 197 of these general local governments.

Relations among these territorial governments are characterized by many forms of physical, social, and economic specialization and interdependence. For example, center cities serve as corporate and banking headquarters for industries located in urban fringe areas, whose workers reside in still other suburbs. Further specialization results in localities concentrating particular kinds of manufacturing and commerce. For example, high-technology computer industries locate in Silicon Valley, a cluster of towns on the peninsula south of San Francisco.

Suburbs are stratified by fine distinctions of class and income (Markusen, 1976).[4] These patterns of residential specialization reflect deeply rooted beliefs that housing type and tenure mark not only status but also progress through life (Perin, 1977). Moreover, as municipal corporations, cities respond to economic incentives to become or stay as affluent as possible. The net result of many cities specializing in fine gradations of class and income is that some are left with the bulk of the poor.

Although disparate and discriminatory residential and taxing practices tend to exacerbate inequities among these jurisdictional components of the metropolis, issues such as employment, transport, and environmental protection bind them together in common concern. Suburbs specialize in terms of lifestyle, manipulating who may reside there through zoning practices and other mechanisms. But they collaborate on such system-maintenance problems as transit. "Policy areas which are perceived as neutral with respect to controlling social access may be centralized; policies which are perceived as controlling social access will remain decentralized" (Williams, 1971, p. 93).

Specialization breeds spillovers. Elaborate commuting patterns caused by highly specialized residential and industrial areas in turn cause air pollution, which also flows over political boundaries. Even seemingly autonomous local activities such as police protection and garbage disposal acquire "externalities" in these increasingly interdependent spatial networks.

Spillovers spawn numerous, proliferating special purpose districts (Bollens, 1961), which vary in scope and power from mosquito abatement districts to port authorities. Special purpose districts respond to natural, physical circumstances, such as bays, that transcend local political boundaries. Special districts also respond to problems caused by increasing specialization and interdependence: externalities and concerns with efficiency and economies of scale.[5]

Territorial political structure further complicates institutional arrangements in a metropolis. Congressional districts are different from state representative districts, which are in turn different from city and county boundaries. These differential geographic constituencies create representational overlap (Landau, 1969), which provides citizens with multiple channels of political influence. At the same time, overlap precludes hierarchical political structure and deters fixed political allegiances.

Scholars and public servants debate the value of this complex, multicentered system of area governance. It is not merely fragmented, critics say, it is fractious. Moreover, it is irrational, inefficient, and unfair. In its place, they propose regional government, which would treat the metropolis as a whole, with technical economies, noncompetitive comprehensive planning, and a redistribution of taxes and services that is equitable. On the other side, proponents of the messy, specialized, metropolitan network cite the previously noted benefits of multicentered federalism, such as multiple cracks, as well as the superior responsiveness attributed to small government. Some economists go a step further to conclude that the overlapping specialization of multiple governments is efficient (Ostrom, Tiebout, & Warren, 1961). Indeed, Tiebout (1956) says that this constellation of specialized governments is no less than a market of baskets of public services for given tax prices.

Area government, a key dimension of the intergovernmental system, is complex. Instead of the self-sustaining, Jeffersonian, traditional community of common concerns fashioning government to suit its own unique interests, area government has become an elaborate and specialized system of governments. General purpose governments are stratified by type of commerce, industry (if any), income, and lifestyle. But such fine-honed distinctions do not create modern communities of interest. People's interests and public problems transcend these general government boundaries.

THE TIME DIMENSION

Time is a key dimension of any analysis of development and dynamics within a system. Time is also central to the study of intergovernmental relations; because the governmental units are seen as acting with respect to one another, short of isolation, their relations cannot be understood as static. The intergovernmental system thus entails processes.

Analysis over time also helps one understand present dynamics, because previous intergovernmental program structures may be powerful determinants of feasible successor programs, and past agreements and disputes tend to influence bargaining in the present.

Although sectors usually have latitude in their choice of strategy, technical demands impose their own time constraints. Development of a single energy plant currently requires a 12-year lead time. Other sectors may move quickly, depending on their policy choices and technological capacities. Swift implementation of the Salk polio vaccine program virtually eradicated a dreaded disease on a national scale in a short time.

Different time horizons add to the complexity of intergovernmental decision making. For example, an areawide planning organization may be dominated by the perspective of its 20-year land use plan, whereas a low-income suburb may be enticing industry to alleviate an immediate fiscal crisis, and an environmental agency may be administering pollution standards. This scenario suggests prospects for lively interaction among different sectors and levels of government. It also illustrates the importance of different tasks' time frames: long-term future, immediate crisis, and day-to-day equitable administration.

Finally, time itself shapes the processes and form of intergovernmental relations. For example, short-term, ad hoc intergovernmental arrangements are apt to be more flexible and so more adaptable to change than longer-term, more permanent legislated intergovernmental arrangements, which are liable to over-specify and rigidify authorities and responsibilities.

In sum, time is an indispensable dimension for analysis of intergovernmental dynamics. Together with the vertical and horizontal dimensions, it promises a way of ordering the apparently confusing relations among governmental agencies.

THE DECISION MODE DIMENSION

The fourth dimension, decision mode, is not embedded in theories of federalism. Yet, because actions among governments imply decision making within and among governmental organizations, the way they reach those decisions seems a critical factor affecting intergovernmental relations. A second reason for including decision mode here is to free analysis from misconceptions and clear the way for observing how actual public choices are made in complex intergovernmental systems.

A third reason is to explore how decision modes vary. Different decision modes imply different types of intergovernmental communications and exchange. In some sets of intergovernmental relations, officials may collaborate fluently as peers, whereas in others, they may be speaking different languages. The following sections discuss differing decision modes according to their majority and interactive aspects. Decision modes influence how agencies interact, and thus help shape the dynamics of the intergovernmental system.

Majority

A popular misconception is that U.S. government operates by majority rule. Pure, simple majority decisions rarely occur; a Vermont town meeting offers the exception that proves the rule. The various forms of fragmentation, specialization, redundancy, and representation discussed above, which have been designed into and evolved from the U.S. governmental system, prevent simple majority rule. Indeed, the essence of federalism, the division of power, precludes a single, encompassing majority rule.

First, the intergovernmental system comprises many subcomponents—committees, councils, legislatures—governed by majority rule. The system contains no majority rule for the whole. On no occasion—not even a presidential election—does the majority of the populace's electorate determine the outcome. The representational designs of Congress, states, and most municipal legislatures are republican, intentionally removing public decisions from popular majority rule.[6] Public choices are reached by majorities of elected representatives through a series of voting mechanisms. Rules that depend on majorities protect against rash popular judgments. For example, the constitutional amendment process is deliberately ponderous to ensure stability of the rules themselves.

The whole system's extra protection through extra majorities, however, is complemented by majority rules that in effect protect minorities.[7] Subdivision permits minorities of the whole to be majorities of smaller jurisdictions. Cities may impose their minority preferences on the majority of a metropolitan area. Parochial, and sometimes discriminatory, growth, zoning, and taxing practices of some jurisdictions may result in inequities that the popular majority of the larger metropolitan area would never condone, much less vote for.

Second, the bulk of public decision making is never put to popular vote. Rather, through representational arrangements and administrative delegations, much public choice is conducted through committees and agencies that operate by other decision rules. Although deliberations may end in a formal vote, the real work of making a choice occurs through committee members or agency staff discussing, bargaining, and building consensus.

Moreover, as functional federalism suggests, committee and agency decisions are linked to special or private interests sheltered from popular vote. The idea of interest groups having such influence is

> intended to balance the inherent bias of democracy in favor of actions that benefit the majority a little even when they hurt a small minority a great deal. Yet these groups seem to tilt the balance in the other direction, often obtaining benefits for the relatively few they represent at the expense of the unorganized majority. Their power is enhanced by the costliness of information about the legislative process. (Okun, 1975, p. 28)

In sum, voting through majority rule is less important to public decision making than alternative decision modes.

Interaction in Deciding Processes

The extent of interaction among decision makers also shapes the intergovernmental system. Interaction is considered by four degrees: isolation, bargaining, consensus building, and social construction of reality. Table 4.1 offers a loose guide to the following distinctions.

Isolated, autonomous registering of individual preferences is the presumed basis of our political method of voting. Voting and the necessary polarization of choice emphasize winning and losing. Secret ballot voting protects individual freedom of choice not only in the present but also in the future. The voter is beholden to no one and incurs no obligations to vote the same way in the future.

Bargaining entails more interaction. Although preferences are distinct, participants may debate and trade to reach decisions that provide

Dimensions of the Intergovernmental System

TABLE 4.1 Extent of Interaction in Deciding Processes

	Deciding Processes			
Characteristics	Isolated Voting	Bargaining	Consensus Building	Social Construction of Reality
treatment of conflict (multiple preferences)	emphasizes conflict; win-lose outcomes	creates win-win possibilities from conflict	submerges conflict	eliminates conflict
future impacts	none	precedents obligations	mutual commitments	continuing reference groups
rule	scrupulous procedural rules to provide legitimacy	processes variable but understood in advance	flexible	tacit group norms, almost subconscious
decision-group conditions	large numbers, diverse status	small numbers; minimal trust often unequal status	small numbers; considerable trust, appearance of equality	small numbers; group identification; peers
benefits	competing ideas	accommodation of difference	common goals	deep agreement
dangers	polarization	collusion	obscured, suppressed issues	no consideration of alternatives

every party with some benefits. Negotiations may reduce conflicts among preferences and pave the way for less divisive decision making in the future. Whether cordial or hostile, bargains tend to set the tone, precedents, and obligations for subsequent negotiations.

Consensus building involves more interaction to achieve a collective, shared benefit, rather than the more separate quid pro quo trades of bargaining. Differences are submerged to generate convergence toward common interests. Discussion may reveal new, superior choices that none of the participants could have recognized before the collective effort. Like bargaining, consensus building influences future decisions and tends to bind participants together more closely in mutual commitments.

Social construction of reality depends entirely on interaction. Participants do not even perceive different preferences or opinions. The question is not a matter of one's values pitted against another's. Rather, something is at issue and needs to be sorted out. Participants collectively make sense of uncertain, confusing events and conditions (hence the term *social construction of reality*). Their collective decision derives more or less automatically from their common understanding. In contrast to isolated voting, where status is irrelevant, and to bargaining, where differential status can be critical, in social construction of reality, participants must be peers.

To illustrate the process, schoolmates (peers) faced with a derelict in their play area (unusual, confusing event) collectively make sense of his antics by interpreting them as funny or frightening (social construction of reality), and then either laugh and continue playing or run, according to their understanding (common decision deriving automatically from common interpretation). Similarly, engineers faced with an entirely new building product collectively interpret it as useful or as a threat to their professional practice, and then act accordingly. Peers form a group for interpreting situations on a more or less continuing basis.

These different degrees of interaction shape decision processes within and among governmental institutions. Conditions tend to dispose some governmental combinations toward more interactive decision processes and others toward more isolated ones.

The number of participants determines the range of feasible processes. Very large numbers are incapable of direct, thorough exchange of ideas, and so their participation is reduced to registering yes-no, a or b votes. Although the U.S. system appears plagued by large numbers, diverse representational schemes[8] permit more interactive decision processes than isolated voting.

When numbers permit discussion, the extent of interaction depends on the degree of trust among participants. Labor-management arbitration is unlikely to develop shared perception of conditions because of distrust bred through job position, training, and social class differences between the participants. On the other hand, state and local law enforcement officials are likely to construct a shared reality from uncertain conditions in light of their similar training and experiences. Depending on the number of participants and the trust among them, some intergovernmental situations may require voting, whereas others may lend themselves to bargaining, and still others to consensus building.

So far, discussion has implied that the degree of interaction in a deciding process follows from given conditions. Although conditions usually set some constraints, they can be manipulated to influence the extent and style of interaction. Indeed, much of the organization literature stresses participatory processes within organizations;[9] and some planners (Susskind, 1981; Susskind & Ozawa, 1983) advocate consensus-building arrangements (e.g., Godschalk, 1992; Innes, 1996; Susskind & Cruickshank, 1987) to deal with intergovernmental issues. Some states have created innovative arrangements for interactive planning.[10]

If the type of interaction is indeed variable and yields different results, it may be a powerful public policy instrument. At this point, however, the policy potential is largely unexploited. The actual extent of interaction seems to derive more from evolved tradition, informal cooperation (Chisholm, 1989), and political opportunism than from conscious policy design.

Although proposals to design interaction imply that the greater the interaction the better, judgments cannot be so clear. Although increased interaction probably increases communication, it may not increase the breadth of ideas considered. Moreover, it is liable to suppress conflict and thus potentially fruitful debate. On the one extreme, social construction of reality depends on an accepted reference group of peers, which is constituted precisely on the grounds of member sameness. On the other extreme, secret ballot voting creates the conditions that make wide differences in politics, income, and status—at least for the issue at stake—irrelevant.

Power imbalances may skew bargaining and negotiation. Questions of who participates and how they communicate pose difficult trade-offs and prompt efforts at improving understanding between competing groups (Forester, 1989, pp. 137-162). Some interests may feel so excluded by the establishment that they operate outside the political and administrative system and take recourse though the courts. For example, cases brought by people displaced by urban renewal led to new program requirements for relocation assistance. Environmental suits have altered projects and programs (e.g., Howitt, 1984). Stoker (1991, pp. 105-106) says law suits are more likely in initiatives based on legal rights, such as the guarantee of an appropriate public education for all children, regardless of disability. On the other hand, the threat of costly legal battles from citizens is prompting the rise of collaborative and consensual processes in interagency cooperation (Thomas, 1997).

Interactive aspects of decision modes suggest new ways of understanding the intergovernmental system. For example, similarities in decision rules permit some kinds of intergovernmental collaboration. Some sets of agencies may be so accustomed to working together that they collectively can perceive problems in only one light and so collectively can devise only one kind of solution.

SUMMARY

To build an analytic framework, this chapter has extracted key dimensions structuring the U.S. intergovernmental system: vertical, horizontal, and time. On the vertical sectoral dimension, government agencies align themselves in interdependent chains linked by specialization. On the horizontal territorial dimension, governmental units specialize as well. Time affects intergovernmental relations and may disclose interactive patterns. A fourth dimension, decision mode, is included because of its potential power to explain how multiple governmental units interact.

When variable decision rules are considered in conjunction with the multiple time frames, the functional chains of agencies, and area governmental units, the resultant complexity no longer seems puzzling. Indeed, Kiel (1994) shows different patterns of chaos operating even in a single government agency. As in the chemical world, the individual elements of the intergovernmental system combine at different rates according to different rules into different types of clusters to yield a rich and varied system. These four dimensions dispel the illusion of simple government.

The next chapter uses these dimensions to sort intergovernmental exchanges, to show recurring patterns, and to disclose the operating structure of the intergovernmental system.

NOTES

1. Clearly, in a very broad view, patterns and causes exist. The hierarchical arrangement in each policy concern is largely shaped by its intergovernmental financing arrangement. In particular, national grant-in-aid programs have been partly responsible for the increased centralization in U.S. governance. Periodic partisan political efforts to dismantle or devolve grants-in-aid are proposed in terms of federalism. Nonetheless, particular

Dimensions of the Intergovernmental System

intergovernmental arrangements emerge from the interaction of particular historical and political circumstances seemingly immune to technological dictates or larger causal patterns.

2. Many other terms for sector convey the same general idea of multiple organizations linked together in the same policy area. Milward (1982) provides an early background literature and description of interorganizational policy systems (sectors). In functional fields (or functional specializations), some have seen "a growing trend toward 'sectorization'" with services being defined in limited, functionally differentiated terms, for example housing, employment, child care, and mental health (see Wildavsky, 1979; Burstein, 1991) (Scott, 1992, p. 222). Mandell (1990, p. 35) finds that public interorganizational networks have been referred to as *interorganizational policy systems* (Milward, 1982), *mandated networks* (Raelin, 1980), *implementation structures* (Hjern & Porter, 1981), *issue networks* (Heclo, 1977), *management networks* (Wright, 1983), *directed interorganizational systems* (Lawless, 1981), and *federations* (Provan, 1983). Still other terms (e.g., *implementation networks*—O'Toole, 1993; *multiorganizational systems*—Chisolm, 1989) refer to the same general concept.

3. The sectors are institutionalized in the sense of "the persistence of practices in their taken for granted quality and their reproduction in structures that are to some extent self-sustaining" (Powell & DiMaggio, 1991, p. 9). The divisions into sectors are institutionalized and accordingly understood as "natural," in that the distinctions are widely believed even though they cannot be proved. Sectors seem natural distinctions of public policy. Similarly, "knowledge is seamless, but conventional distinctions have developed between various scientific and humanistic disciplines with the result that most colleges and universities exhibit the same or very similar collections of schools and departments" (Scott, 1992, p. 209). Thus, the division of sectors into housing, transportation, health care, and so forth appears to be obvious.

In reality, however, the sectors impinge on each other at the margins. These peripheral connections (planned, unplanned, and ignored) and their implications for policy and the operations of the intergovernmental system are explored in following chapters.

4. Although Markusen (1976) and others have shown this fine demographic specialization among suburbs, it is important to dispel the prevailing myth of abject center-city poverty ringed by suburban affluence. A 1979 study of six metropolitan areas showed that the suburban share of all very low-income households (defined as having incomes below 50% of the Standard Metropolitan Statistical Area median income) in each metropolitan area was surprisingly high: ranging from 42% to 67% (Christensen & Teitz, 1980, pp. 185-202).

5. Separating public consumption from public production of a public good or service can increase efficiency (Ostrom, Tiebout, & Warren, 1961). Many smaller general local jurisdictions can consume benefits produced by larger special districts. For example, the Northern California-East Bay Municipal Utilities District serves 16 localities.

6. Direct, popular initiative processes, such as those in California, appear to counter the caution inherent in republican government. Yet few of the initiatives passed are implemented without modification by the more ponderous republican process and often with intervention through the courts as well.

7. Even Madison, a proponent of strong union, was concerned with protecting individual liberty, and so may have intended this effect: "Extend the sphere and you take in a greater variety of parties and interests; you make it less probable that a majority of the whole will have a common motive to invade the rights of other citizens" (Madison et al., 1937/1787-1788, p. 62).

8. The familiar form is the traditional representation by location—by neighborhood district, by state assembly district, by congressional district, as well as by city, state, and nation. Less familiar forms stem from interest, for example, the National Rifle Association, with association officials and lobbyists representing the full membership. Broader, more powerful interests, for example, labor and agriculture, are directly aligned with sectors. General government demands representation and so forms associations called "public interest groups." Thus, for example, cities band together to create the League of Cities, with officers and lobbyists representing the full membership. Similarly, a committee of six state governors represented all states in bargaining with the Reagan administration over new federalism proposals.

9. A school of advocates of participatory, rather than bureaucratic, organizations grew out of the human relations school, which originated in the classic studies summarized in Mayo et al. (1951). A representative book advocating participation in firm decision making is Argyris (1964).

10. For example, Florida, Maine, Vermont, Rhode Island, Georgia, Washington, and New Jersey have required consistency between local and/or regional, state, and contiguous jurisdictions (Gale, 1992, p. 430). In practice, New Jersey, Vermont, and Georgia planners engage in intergovernmental negotiation (Gale, 1992, p. 434) and cooperative planning (Bollens, 1992, p. 458). O'Toole (1993) reports efforts at structuring the best interorganizational arrangement for a policy, but presents compelling cautions.

CHAPTER 5

SECTORAL DYNAMICS

Institutionalized Technological Elaboration and Effects on Area Interests

Preceding discussion has laid a base for understanding the structure and dynamics of the U.S. intergovernmental system. Its framework is ordered by four key dimensions: vertical (sectoral), horizontal (area), time, and decision mode.

The chapter follows the ordering dimensions, first analyzing interactions among units along the vertical dimension, sectoral dynamics. Second, the chapter analyzes the vertical dimension's juncture with the horizontal dimension, a sector's effect on an area. Third, it analyzes how many different vertically aligned policies affect areas over time. The chapter concludes by assessing the effects of these intergovernmental dynamics.

Figure 5.1 offers a preview of the chapter's major sections.

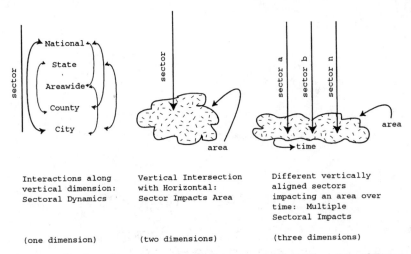

Figure 5.1. Sketches of Interactions Among Governmental Units Along Several Dimensions

SECTORAL DYNAMICS

This section analyzes governmental interactions within the vertical dimension. Assessing patterns of exchange shows that some sectors are fairly stable, whereas others seem in upheaval, with much collaboration and conflict. Interactions along the vertical dimension reflect development of a sector's methods.

A sector is a vertically linked chain of federal, state, areawide, and local agencies addressing a common functional specialization through interdependent tasks and procedures. The specialization constitutes the mission or *mega goal*. In the housing sector, it may be "more and better housing"; in health, "more and better health care"; in crime, "less crime." The common goal[1] in a sector precludes disputes over the ultimate purpose and concentrates controversy on ways to achieve it and what it means operationally. For example, agencies debate what is water quality.

Arguments and negotiations over how to formulate and achieve a goal imply that interactions within a sector rarely occur through apolitical administration, but rather through a more political process. A controversy over means often harbors conflicting values regarding tasks. All agree on the "more and better" goal, but one housing faction urges the filter-down approach, which serves higher-income people first, whereas another faction favors eliminating the worst conditions first. Variations and innovations in a sector's technology may have redistributional effects (Zysman, 1977). Characterizing interactions in a sector as jockeying over sometimes value-laden means to accomplish a shared mega goal is useful for the following analysis.

A sector's set of subdivided tasks constitutes its task technology. This book uses the term *technology* in its broadest meaning, the knowledge of how to do something. Technology is the instrumental means by which people, tasks, procedures, and materials are put together to accomplish something.

This public version of an economist's production function may be automatic or mechanical. For some public tasks, technology is derived from sound science. The Salk vaccine, for example, protects children from polio. Sanitation protects against cholera.

But for many problems, task technology is far from straightforward. Complications can arise over the choice of technology, the reliability of a chosen technology, or the precise operational definition of the goal. Thus, technology may or may not derive automatically from sound science. Even in the case of vaccinating children, although biochemistry remains exactly the same, operational technology has shifted from a short-term, intensive, school-administered vaccination to a dispersed, decentralized, physician-administered vaccination.

In the case of housing, some would say the technology is well-known and reliable: the same wood-beam traditional method used by carpenters for many years. Others would say that method is precisely the public problem; private, traditional housing construction and finance now cost more than many people can afford. Thus, technology conveys not merely the mechanics of achievement but achievement within an implicit or explicit set of standards and constraints. Technology, the knowledge of how to do something, depends not only on reliable knowledge but also on public acceptability.

Technology is a way to achieve a political goal. Thus, an effective technology has proven itself both workable and acceptable. When technology is unclear, it may be technically unreliable, insufficient to the goal's constraints, or never have been tried.

People working in multiple agencies in the same sector share not only their common goal and interdependent, subdivided technology but also a common orientation. Their shared sectoral perspective on public problems derives from their work experiences and the programmatic tools and skills they have available. For example, the engineering profession dominates the transportation sector. The common orientation ensures that issues are cast in the same mold. Common assumptions behind the sectoral orientation are tacit and, accordingly, never debated. Consequently, sectoral issues are narrowed to technological variations.

Chisholm (1989) observed interactions in the transportation sector and found coordination through informal channels. He concludes that "a multiorganizational system whose component organizations were characterized by similar technologies, common professional backgrounds and generally similar goals facilitates the development and maintenance of informal mechanisms" (p. 193), which then enable coordination through mutual adjustment.

This strong common sectoral orientation, and often professional discipline as well, fits with the civil service. In combination, this stable career structure builds a cadre of experts who have worked together for many years. They meet repeatedly over projects and problems and in professional associations. Often they change levels of government and even exchange jobs. In the course of sharing experiences, as well as the goals and orientations of their sector, they form a reference group. Through their specialized social construction of reality, they develop and refine their sectoral technology.

Thus far, the sectors have been depicted as chains of agencies, linked vertically to contrast them with territory-based government, which is seen on the horizontal dimension. Agencies in a sector link vertically up and down federal, state, areawide, and local levels, united by their functional specialization.

Sectors have become still more complex as they cope with stress and, furthermore, as interest groups have inserted themselves into the intergovernmental fray. Thus, at each governmental level, several governmental agencies and associated nonprofit and for-profit organizations may operate in the same sector. Chisholm's (1989) study, for example, involved transportation agencies in several counties in the San Francisco Bay Area. Prompted by the 1990 Affordable Housing Act and

Sectoral Dynamics

continued funding cuts for affordable housing, nonprofit housing developers have grown and are supported in a variety of ways by local government and nonprofit intermediary and service organizations. (See "A Day in the Life of a Planner in the Housing Sector" in Chapter 4, "Dimensions Structuring the Intergovernmental System.")

Agranoff and Rinkle (1986) describe similar complex intergovernmental decision making in the human services sector. They note that "while the array [of grants, responsibilities, auspices] has become more diverse, it has also become more interdependent with coordination requirements, multiple funding sources, matching requirements, and most important, the need for providers in one jurisdiction to secure services from another." Indeed, the increasing complexity makes for interlocal and interregional collaboration within the sectors. The relations may appear fragmented, but they encourage combinations and recombinations among the agencies. The increased complexity and increased roles for state and local decision making constitute further elaborations of the sectors. The important organizing principle of the sectors is their shared functional specialization.

Because of their agencies' interdependence, a new finding, disaster, success, or political demand at any level triggers reactions throughout all levels in the sector. In housing, for example, national-level officials may discover that plastic pipe is just as effective,[2] easier to use, and less expensive than lead pipe, and so adjust their subsidized housing procedures to promote plastics. But, because local housing codes require lead pipe (at the insistence of strong local labor unions), either local officials must adapt their codes or national officials must abandon their idea for reducing housing costs, or nationally subsidized housing will not be built in that locality.

To deal with this stress, interdependent housing agencies may take a range of actions described generically in Chapter 3, "Interactions Among Governmental Institutions." The agencies may collaborate by making minor adaptations in their procedures or forming an intergovernmental task force. If, instead, these agencies deal with the stress through conflict, local institutions may rebut national findings with contradictory data (e.g., on plastics' carcinogenic characteristics) or counter with an alternative program (e.g., housing funded jointly by city bonds and labor unions). Whether the result of collaboration or conflict, the new procedures, task force, data, and programs all focus on methods and task division. The various interactive responses all address the same general substantive concern, in this case, housing.

In sum, interactions among institutions within sectors are stimulated by upsets to their existing technology and interdependent task subdivision. Sectors' piecemeal (Hirschmann & Lindblom, 1962, pp. 211-222), problem-oriented (Cyert & March, 1963) solutions are close to usual practices, in part because such minor adjustments are easier and safer to adopt than a major change. Moreover, common goals, problem perspective, training, and work experiences bind sectoral experts to their familiar methods. Rather than consider fundamental change, they refine their existing technology.

Interaction among institutions may seem perpetual and chaotic. In housing, for example, a program's authority is centralized, then decentralized, then passed on to the states, while strategy shifts from construction industry subsidies to experimental mass production to neighborhood preservation to consumer subsidies and then to nonprofit housing development. Other sectors, such as health care, go through similar upheavals. Through various collaborative and divisive interactions, agencies at multiple levels alter their organizational designs and procedures to generate and accommodate such technological change.

Other sectors, however, experience little disruption. For example, for some time, traffic engineering has had stable arrangements. (In congested urban areas, however, the sector is experiencing turbulence and, accordingly, exploring new procedures such as congestion pricing.) Stable sectors have no need for reconstituting strategies, organizations, interorganizational relations, task subdivision, and procedures because their existing technology is effective. Reformers, whether politicians in search of campaign issues or troubled academics and professionals, bypass proven methods.

On the other hand, reformers are likely to criticize, research, write exposes, propose radical alternatives, and threaten budget cuts for ineffective sectors. In response to such buffeting, sectors are apt to modify their technology, partly to appease critics and regenerate support through a new strategy and partly in their ongoing search for something that will work. Then, too, because the technology is immature, informative surprises—both disastrous, unforeseen consequences and accidental successes—frequently emerge from putting the technology into practice. These surprising results, as well as internal and external criticisms, stimulate the chains of interaction described above. These interactions constitute an institutional search for an effective technology.

The degree of change in the revised task technology—often understood to be "the answer" to whatever problem prompted the revision—ranges from minor adjustments to major transformations. The extent of change depends on the severity of the substantive challenges and the political forces driving the change. Changes occur frequently, but may go unnoticed or be widely publicized, depending on the political and media attention directed at the changes. Minor adjustments happen constantly. Major upheavals, such as revenue sharing, numerous defense base closings, and the decision to end welfare, tend to be episodic.

As Kingdon (1984) explains, major policy change takes place when a policy stream connects with the political stream and a window of opportunity arises. Program ideas develop in the policy stream, often over a long period of work by academics, congressional committee staff, and senior civil servants. These people are key members of the sector who, as they are reformulating policy, are developing the sector's task technology. According to Kingdon, entrepreneurs (often lobbyists in the sector) link the policy and political streams, making the connection between the salience of the policy reformulation and the exigencies and opportunities of the political moment.

Major reform develops in the sector but requires connection to political demands and opportunities to be adopted. For example, the idea of vouchers to enable poor people to choose housing in the private market emerged from the housing policy stream and was about to undergo a rigorous experimental test in the early 1970s (RAND Corporation, 1982). At the same time, the voucher idea swept through a policy window into the 1974 Housing and Community Development Act and became nationally available, used, and popular long before the formal experiment proved the idea to be feasible. For another example, health care policy specialists had identified the need and proposals for reform decades before the political stream was ready to address health care reform.

Much elaboration of task technology occurs outside the political spotlight, however. Policy change is still prompted by surprises and political exigencies; the problems simply pose lower stakes, or at least are considered so by the major political actors and the media. At the midrange level, policy changes are developed in the sector's policy stream (e.g., committee staff and senior civil servants) and carried out through amendments to existing legislation. For example, the formula distributing community development block grant funds to cities has been repeatedly revised. Such reformulations occur frequently, without fanfare.

Similarly, less significant policy and program changes often occur through changes in regulations. Again, the changes are prompted by surprises and inconveniences, both political and technical, that arise in the sector's task technology. Problems from applications elicit reforms that, although substantive, are not significant enough to require legislative change.

People working in the sector's agencies follow the changes, which they themselves prompt, and adapt their behavior to the changes. They have ways of contributing to and following the changes, for example, through the *Federal Register* and specialized associations' participation and publications. Publications such as the *Housing and Development Reporter* (Bureau of National Affairs) carry the day-to-day revisions.

Even on a minor scale, program officials at all governmental levels engage in changes to adapt policies and programs to local conditions. Agranoff and Rinkle (1986) show that such successful reforms occur through interactive problem solving within the services sector and are politically sensitive to local conditions. Implementation of national acts brings adjustments to particular places' political and physical conditions. For example, the regional office of the Environmental Protection Agency (EPA) defined water quality in terms of water quantity to address the issue of salinity levels in the San Francisco Bay.

At every step, multiple actors are involved in the sector's reformulation of its task technology. Political officials, whether at local, state, or national levels, participate to enact changes, for example, using political channels to generate pressure or resist a mandate to protect a constituency. Sometimes they also create a policy window, for example, through campaign issues. Sector specialists, such as academics, consultant evaluators, legislative staff, and senior civil servants at every level, contribute program knowledge and specialized expertise. Lobbyists for various interest groups in the sector help connect the policy and political streams, operating at national and subnational levels. For example, the National League of Cities, which works on behalf of cities' interests, also has a California League of Cities, and the American Medical Association has its state counterpart, California Physician.

This specialized, political, and intergovernmental process of policy reformulation may also redistribute power and resources. Even on a small scale, changes in rules and procedures may alter incentives. Midrange changes redistribute power and authority among levels of governments. For example, some reformulations may curtail cities' options. Major reformulations affect the intergovernmental actors and often private and nonprofit actors affiliated with the sector as well. Partisan political rhetoric such as the recent "Contract with America" may amplify the redistributional shifts, for example, to promote transfers to the private sector, or the rich, or local authorities.

In short, agencies within a sector share a common goal that they jointly address through subdivided, interdependent tasks that constitute the sector's technology. They elaborate their technology directly through joint problem solving and indirectly through adjusting reactions to each other and to outside pressures. Generally, the more immature a sector's technology, the more turbulent its interagency relations; the more mature a sector's technology, the more stable and ordered its interagency relations. Over time, the interactions in a turbulent sector constitute a complex search for an effective technology.

Sectors Distinct From Each Other

Sectors tend to be quite distinct from each other.[3] Sectors act separately from one another in part because of their ties to private interests. Consequently, developing interagency coordination between the sectors (in contrast to the ease of coordination between agencies in the same sector) proves difficult. Benson (1982) finds a problem of "interorganizational feudalism" because of historically weak central government in the United States that cannot overcome the vested interests that dominate the sectors. Lowi (1969) shows how governmental agencies work with vested interests.

Proliferation

Sectors are not monopolies; nothing in the intergovernmental system prevents multiple approaches to public policy. The intergovernmental system encourages sectors to proliferate in four connected ways.

First, agencies and sectors multiply to accommodate multiple interests. Second, agencies make decisions in a piecemeal fashion (Hirschmann & Lindblom, 1962), acting in their own interests with respect to other governmental agencies. They form each other's environment, posing problems and opportunities for each other. The agencies' interaction occurs through partisan mutual adjustment. "Intentionally and accidentally neglected consequences of chosen policies will often be attended to either as a remedial next step of the original policy making or by some other policy-making group whose interests are affected" (Hirschmann & Lindblom, 1962, p. 221). Thus, the new programs, legislation, governmental agencies, and sectors that develop in response to pluralistic demands often signify neglected consequences of piecemeal decision making.

A third, related multiplier of sectors is specialization. Demands to remedy the harms caused by piecemeal decision making call for new tactics. So, for example, various improvements in welfare programs created disincentives to work and seemed, relatively, to harm the working poor. In response, the problem was redefined and subdivided. On the one hand, the situation demanded reforms to encourage welfare recipients to work. On the other hand, the situation demanded rewards for the working poor.

This process shows the specialization of poverty. Although the old problem was redefined and sectoral approaches were specialized, the old problems remained intact. Some categories of nonworking poor people (elderly and people with disabilities) nonetheless remained deserving and exempt from workfare reforms. This example illustrates the expansive complexity of specialization. A new finding or political demand prompts problem redefinition, but that new, more complete formulation may not replace an old one. Instead, it expands both understanding and programs.

A fourth source of sectoral proliferation reflects the tension between specialization and integration. The more parts of a process, a program, or sector are specialized, the greater the demand for coordinating procedures and committees (Lawrence & Lorsch, 1967). The demands come from technical requirements, managerial desires for control, and occasional political pressures. In health care, this cycle of conflict between medical professionals and bureaucratic rationalizers (Alford, 1975) proliferates programs and organizations in relation to each other, one asserting the preeminence of its expert knowledge, the other asserting the preeminence of public control. More specialized treatment programs spawn more coordinating mechanisms.

To recapitulate, sectoral dynamics constitute piecemeal, problem-oriented searches for effective technology.[4] Sectors proliferate because they respond to changing and diverse demands and correct neglected side effects of their immature technology. The ensuing specialization induces more programs and more coordinating mechanisms. Because this evolutionary expansion is not checked by the disappearance[5] of earlier sets of agencies (Kaufman, 1977), their numbers increase.

SIGNIFICANCE OF THE AGGREGATE PATTERN OF SECTORAL DYNAMICS

Interactions among agencies within sectors are mutual adjustment processes that collectively constitute piecemeal problem solving. Each new version of a sector's technology is institutionalized into a new division of tasks, responsibilities, authorities, and procedures among the units. In the aggregate, these institutionalized arrangements—now understood as technological elaboration within each sector—are by-products of piecemeal problem solving.

This proliferation of specialized government programs and agencies flies in the face of persistent politically supported efforts to reduce government. Therefore, this growth must serve other functions (Merton, 1957). As noted above, programs and sectors respond to special interests and serve symbolic and political functions. Moreover, the process develops and refines a sector's technology and can thus be seen as an unintended form of learning. The veneer of formal bureaucracy imbues this process with a pseudo-permanence that disguises its trial-and-error form of learning. Indeed, the new institutionalists find that expectations of rationalized bureaucracies make them more prevalent and provide them legitimacy (DiMaggio & Powell, 1983; Meyer & Rowan, 1977). Thus, governmental expansion persists even though unwanted. Driven by internal and external demands for more effective technologies, sectors perpetuate and institutionalize their search for something that works.

Intergovernmental collaboration and conflict appear to revolve around resource dependency. In a sector, resources may be understood as shared, because the agencies' funding and authorities jointly create their sector's goods and services, for example, housing. The resources include funding (grants and related tax and fee revenue arrangements) and authorities. Agencies within a sector jockey among themselves, trying to leverage more resources and fewer constraints for their respective agencies. In that sense, they compete. Nevertheless, over time their interdependence breeds reciprocal support. The relatively small numbers, the common resource pool, the reciprocal relations, and agreements worked out all characterize Ostrom's (1990) collaborative arrangements to manage a common resource pool.

Intergovernmental conflict tends to occur when intergovernmental actors compete not for an increased portion of shared resources but rather for externally controlled resources. Typically, cities compete for industries, which provide the scarce and valued resources of jobs and tax revenues. Competition is intense, with marketing groups selling economic development programs, partly on the grounds that cities and regions must have their own program to survive against rival programs ("Selling Growth," 1995). Poor cities even compete to attract activities usually understood as undesirable, such as prisons and waste treatment facilities (Muller, 1994). The competition includes not only rival economic development and undercutting bids but lawsuits over facilities moving to another town.

In sum, it seems likely that government agencies that depend on shared resources (most often in sectors, but sometimes in places, e.g., watersheds) collaborate. In contrast, government agencies that depend on external resources (most often territorial governments, but sometimes in sectors) compete.[6]

SECTORAL EFFECTS ON AREA GOVERNMENTS

Sectors' searches for effective technologies occur not as abstract thought experiments but rather through real-world applications. This section explores this crucial juncture, where sectoral policy hits the ground of reality.

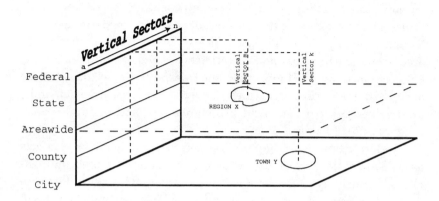

Figure 5.2. Sectoral Effects on Areas

This meeting of the vertical and horizontal dimensions is depicted in Figure 5.2. This section begins by describing how policies affect areas in different ways. Next, it describes intergovernmental grants, a key funding conduit for transforming sectoral policy into action. Finally, it notes how area governments tend to treat a sector's policy and illustrates the juncture of policy and place.

Intended and Unintended Differential Effects

Before viewing sector effects from an area's perspective, it is important to recognize that sectors affect area governments in different ways. Some sector activities affect everyone equally, and so have no particular, place-related effects. Programs such as Social Security serve individuals regardless of their location. Thus, a 65-year-old man who worked 40 years and had one dependent in rural Mississippi would receive the same benefits as an identically described man in New York City.

Other programs affect different areas in different ways. Some involve discretion rather than following entitlement formulae. Some yield benefits or services, such as mosquito abatement, that are necessarily consumed collectively or tied to place. Differential treatment arises from four policy types.

First, some programs serve particular geographic areas. For example, the Tennessee Valley Authority serves a single region. Similarly, empowerment programs, such as some Economic Development Administration programs, focus on especially impoverished communities.

Second, many sectors pursue policies that apply to all geographic areas but adjust to accommodate varied needs and conditions in different geographic areas. For example, the transportation sector tailors its action to the physical peculiarities, demographics, industrial demands, journey-to-work patterns, land use regulations, and political preferences of different areas.

Third, some programs and policies generate differential area effects as investment side effects. To offer a range of examples: Outer continental shelf oil drilling affects communities on the coast but not those in the interior. If a liquefied natural gas installation is sited in one town, it will not be sited in alternative ones. Limited funds mean only a few areas will receive major water resource projects. These cases show differential area effects from place-specific major investments. Unequal treatment of area governments is not the intent, but rather the by-product, of the sectoral investment.

These side effects may be politically exploited. When area effects are considered destructive (e.g., nuclear energy plants), political forces shunt the noxious project to weak or unfavored areas. When effects seem beneficial (e.g., water projects), they may be distributed as political incentives and rewards (traditional "pork barrel"). Some effects generate regional benefits but local costs, such as a waste ground fill, known as a LULU (locally unwanted land use; Bollens, 1992; Popper, 1987).

The fourth source of differential area effect is not merely incidental but undesired. For example, the community development block grant formula of 1974 was designed to treat cities equally[7] but required revision to account for significantly older housing stocks in Northeast cities. Before this modification, the Southwest region was inadvertently favored over the Northeast.

The causes of differential government effects on geographic areas—intended, accommodating, investment side effects (including political pork barrel), and inadvertence—are both understandable and acceptable. Unequal treatment, rather than being a public scandal, emerges from different types of policies affecting communities with different physical, social, economic, and political conditions. Even the strongest critic would hardly expect every state to receive identically designed and priced dams.

The Intergovernmental Grant-in-Aid System

The intergovernmental grant-in-aid system attempts to reconcile interpersonal and interjurisdictional equality with an array of funding mechanisms. The forms grants take vary in the emphasis placed on these general aims and their specific substantive purpose. So, for example, Social Security and unemployment insurance emphasize interpersonal equality. The community development block grant, on the other hand, emphasizes interjurisdictional equality. Tax credits for housing insulation ignore both general aims to focus on the specific substantive purpose, energy savings. Most grant-in-aid mechanisms reflect a more complicated amalgam of aims in a more complicated intergovernmental division of tasks than these "pure" examples.

In the grant-in-aid system, the national government gives funds to state, areawide, and local governments to further a particular policy. Thus, grants directly use and reinforce sectoral alignments (Yin, 1979, p. 17). Some grants serve broad purposes, often block grants such as community development, and give local agencies wide discretion. Other grants focus on narrow, specific categorical tasks such as learning disabilities. Whatever their scope, intergovernmental grants concentrate on means to address their specialty while tailoring particular projects to particular places. Thus, intergovernmental grants make the funding line for a sector's juncture with territory.

Territorial Treatment of Sector Policy

An area government tends to treat a sector through its specialized agency in that sector. A city's public works department deals with the water quality sector, its roads and traffic engineering department deals with the transportation sector, its housing authority deals with the housing sector, its police department deals with the law enforcement sector, and so on, specialty by specialty. Such delegation of labor seems to make sense because government's work is too much for any one person or team of generalists, and the specialists are, comparatively, experts. Meanwhile, the locality is likely to be competing with neighboring communities for better tax bases and bond ratings, and so these specialists tend to be alert to opportunities.

In the heyday of national funding, these specialists practiced grants writing, the art of securing the sector's money for their community. Even after national (and some key state) funding cutbacks, the role of the local specialist remains the same: to promote the sector at least cost to the local government. It is in mayors' interests to delegate issues to specialists. It is in specialists' interests to promote their respective specialties.

For example, the city of Boston recruited the nation's top urban renewal expert. He then secured preliminary grants and began to build the Boston Redevelopment Agency (BRA), then the local branch of the community development sector. The BRA created plans to raze the West End, using U.S. Department of Housing and Urban Development (HUD) money, for the renewal and greater glory of both Boston and the BRA. The BRA, like redevelopment agencies in some other cities, continues to operate even though HUD's redevelopment funding was folded into the community development block grant.

A sector applies its program through its local unit. The sector tailors its program to those local conditions. Many sectors' programs "hit the ground" in localities. For example, group homes for people with disabilities constitute a program in the mental health sector and are physically located in cities.

Disjointedness of Multiple Sectors Affecting Area Governments

Sectors differ in their decision structures and in their technologies' time frames, maturity, and, accordingly, frequency of modifications. Therefore, sectors collectively affect area jurisdictions in a fashion that is necessarily unpredictable and piecemeal.

Box 5.1

Consider the plight of a typical center city. One day it may be granted a major water project, the next week it may be confronted with radical budget reduction in its urban renewal projects (which have been under development for 15 years), and a year later it may be the site for an energy plant, a consolidation of neighborhood mental health clinics, and community demands for neighborhood economic development. This composite scenario is understandable in terms of interactions within the separate sectors. It illustrates centralized (energy) and decentralized (community development) decision structures, mature (water)

Sectoral Dynamics

> and immature (mental health care) technology, long (urban renewal) and short (clinics) time frames. Sectors proceed on their own momentum, affecting areas haphazardly. Although each sector's action follows from its sectoral logic, from the area's perspective, many sectors' actions seem collectively chaotic. The assorted sectoral effects are so disjointed they are confusing. In the hypothetical example, sudden water-project development is followed by radical urban renewal cutbacks, but the employment disruptions and physical dislocations cannot be coordinated. Similarly, the health sector's shift away from neighborhoods is offset by grassroots economic demands. Countervailing sectoral demands defy concerted city policy.

Because each sector acts according to its own logic and momentum, and because tasks are so intertwined that no single level can rule the others, no one can coordinate the sectors. Therefore, no government—not even the national government—can develop a coherent policy across sectors. Moreover, most states, counties, and cities cannot even predict, much less coordinate, their reactions to the diverse, disjointed sectors' impositions.

Consequently, area responses to sectoral demands tend to be relegated to the special purpose agency associated with the particular sector. The agency, sharing the sector's goals and training of the other agencies in the sector, reacts to its sector's concerns and opportunities. Thus, the local jobs training and placement office responds to the jobs sector's emphasis on private industry participation, whereas the local transit agency responds to transit sector grants for paratransit. Area government tends to respond to sectoral demands from the perspective of the sector rather than from the perspective of the area. Thus, sectoral demands not only affect areas segmentally, they are treated segmentally.

An empirical study of five cities demonstrated that local

> counterpart bureaucracies funded by soft [nonlocal] monies . . . establish close working relationships to the next echelon, e.g., federal area office, . . . within the program, and tend to be isolated from the parallel structures (federal and local) established for other federal aid programs. (Yin, 1979, p. 17)

The study presents evidence on the local components of sectors, as defined and described here, and it concludes that "local affairs may be increasingly tuned to federal regulations and priorities rather than local diversity and agendas" (p. 30).[8]

Area Interests Suppressed by Sectors' Dominance

The nature of area interests is elusive in today's urbanized, physically mobile society, where collective decisions tend to be reached by nationally aligned rather than locally distinct preferences. Specialization, complex interdependence, and easy communication encourage formation of communities that are not geographically bound. Instead, these advanced conditions create a "nonplace urban realm" (Webber, 1964, pp. 108-120), where many different interests are aligned in complex patterns crossing all territorial boundaries. Although some major extractive industries (notably agribusiness, mining, and oil) and some sociocultural patterns (notably Southern traditions) remain geographically based, equating economic and social interests with places is a dubious presumption.[9]

Moreover, the complexity of metropolitan dynamics trivializes the idea of place-based socioeconomic interests. As discussed in Chapter 2, "Competing Theories of the U.S. Intergovernmental System," local jurisdictions now compose a jumble of fragmented but interdependent governments in a collective "real city" metropolis. Whether these socially and income-stratified governments, with their rivalries and artificial boundaries, embody Jeffersonian ideals is doubtful. The question remains whether area governments can or do represent geographically distinguished interests.

The premise of the following discussion is that they cannot, because every area government—even the seemingly most homogeneous—has multiple interests. Constituents have numerous differential interests, including their livelihoods and families, that cross an area's parochial boundary into larger networks. Consequently, representing these diverse interests as if they formed coherent territorially distinct and territorially responsive policy is impossible.

Moreover, representatives consider their locality's future, addressing concerns that transcend current residents' immediate interests. As a municipal corporation, local government also has its own narrow interests, such as maintaining its infrastructure investments, its tax base, its bond rating, and its distinguishing social-economic status. These concerns differ from, and sometimes conflict with, representational responsibilities. Area governments are incapable of representing geographically distinguished interests, not because they fail to pursue area-specific interests, but simply because area interests are multiple and diffuse.

Therefore, these interests cannot be fixed over time. Even fairly homogeneous areas can rarely build a consensus for a general policy that will last. For example, a rich suburb may adopt a popular no-growth policy to protect its open spaces, its tax base, and its racial uniformity. But when a regional environmental plan threatens the policy, consensus is apt to collapse as its original environmental supporters shift allegiance from the suburb's to the larger region's ecology.

Moreover, diverse area interests cannot be ordered by priority. Every area comprises many diffuse preferences jumbled together. Lacking focus, some are brought into relief only by sporadic sectoral demands. For example, a community took a tot lot for granted until it was threatened by plans for a new freeway interchange. Citizens ignored water quality until specialists proposed to add fluoride. A sectoral effect may expose or elicit area preferences that were previously latent, merged with other preferences, or insignificant because they were unchallenged. Collectively, disjointed sectoral demands highlight different area preferences at different times. Therefore, governments cannot respond to disjointed sectoral demands coherently. Instead, areas respond to sectoral demands as they arise, fashioning piecemeal reactions from specialized agencies.

To conclude, reconsider Figures 5.1 and 5.2, which sketch vertically aligned policies affecting an area over time. Sectoral effects on area are determined by sector, rather than by area initiative.[10] Collectively, sectors affect an area in a disjointed fashion. Because area interests are multiple and diffuse, area governments cannot construct coherent area policy. Instead, area governments are left with choices of accepting, adapting, or resisting each sector's programs.

SIGNIFICANCE OF THE PATTERN OF SECTORAL DOMINANCE OVER AREA INTERESTS IN THE AGGREGATE

These disjointed sectoral dynamics generate three troubling consequences for U.S. governance: deterred debate over goals, skewed democratic access, and systemic piecemeal decision making without self-correcting feedback loops.

Deterred Debate Over Goals

The disjointedness of multiple sectors' impositions disguises the reality of diverse, competing interests. The composite pattern tends to preclude pitting one sector's goal against another's or repackaging the various potentially conflicting goals into new forms to suit the area.

This conclusion hardly needs further comment. It goes beyond the system's deterring trade-offs between more butter and fewer guns. Governance is effectively reduced from debating "what ought we to do?" to debating only "how ought we to do these tasks?"[11]

Skewed Democratic Access

Democratic access to government policy is skewed. Public participation is fundamentally reactive rather than formative. In addition, specialization skews democratic access by casting political questions into relatively narrow sector terms and thereby deterring debate over basic issues, such as redistribution. For example, even the current deliberation over a major overhaul of welfare is framed in terms of welfare issues such as how long a family can be on welfare, work requirements, and needs for child care.

General disinterested democracy has lost access, whereas special interest groups (sufficiently educated, motivated, and tied to their specialty) have increased access. The more technical sectors make public participation more difficult, because deliberations may be inaccessible, in sectors' jargon, and incomprehensible. The greater the specialization, the weaker the influence of general opinion (Derthick, 1970). Thus, sectoral dominance over area interests skews democratic access away from the general public toward special interests.

Systemic Piecemeal Decision Making Without Self-Correcting Feedback Loops

In the face of singly formidable and collectively disjointed effects of multiple sectors, area government cannot plan but can only react. Disjointed area reaction to sporadic sectoral demands constitutes systemic piecemeal decision making. Moreover, an area has no prospect of feeding information back to correct problems caused by earlier, disjointed activities. The learning, over time, that emerges from sectors' trial-and-error decision making cannot accrue to the areas they affect. Although sectors may be able to see their trials and errors as short, narrow, reversible, and open, an area's perspective is quite the opposite. A single freeway bisecting a city, for example, has long-term, broad, irreversible, and closed effects (e.g., isolating a ghetto). An area can see the troubling effects but cannot erase them. Even large cities and states, which bear repeated applications of the same sector's evolving technology and so might be presumed to benefit from error correction over time, nonetheless must cope with the enduring and possibly extrapolating (Forrester, 1969)[12] consequences of all the errors. Area governments are left with the debris of sectors' trials.

CONCLUSIONS

U.S. governmental structure can be seen as a frame of vertical sectors and horizontal areas. Its dynamics can be seen as area governments reacting to disjointed applications of sectors' institutionalized searches for effective technologies.

These dynamics breed two kinds of debris. On the vertical (sector) dimension, the debris takes the form of proliferating programs. These are by-products of sectors' piecemeal, institutionalized searches for effective technologies. On the horizontal (area) dimension, the debris takes the form of the lasting and incompletely understood effects of sectors' trials and errors on people and places.

The second major conclusion is that intergovernmental dynamics constrain choice. In today's mobile, diffuse society, area interests are hard to articulate except with respect to sectoral demands. At the same time, sectoral interests obscure area diversity and indigenous agenda (Yin, 1979, p. 30).

Third, this pattern weakens democracy. Discussing grants-in-aid nearly 40 years ago, an expert said, "As long as [state and local] financial deficiencies are not so great as to cost them self-reliance and make them ineffective countervailing influences, I think we need not worry" (Ylvisaker, 1959, p. 39). Yet, the powerful vertical linkages, forged in part by grants-in-aid, have brought about precisely those conditions. Sectors have emerged as rich and powerful factions that area governments cannot counter. At best, they can articulate the costs of sectors' impositions. By funneling democratic access to sectoral reactions, the intergovernmental system negates its early promise of citizens' power to become government rather than merely to resist it.

NOTES

1. To assert that the agencies in a sector share the same mega goal by no means implies that the agencies have no other goals or that the formulations remain constant. O'Toole (1989) points out that even when a chain of agencies shares an agreed goal with a clear definition, agencies still attend to other goals related and unrelated (e.g., overlay statutes such as access for people with disabilities) to the shared goal. Moreover, the mega goal harbors many subgoals, such as innovative technology for achieving clean water, or housing with services for people with special needs for achieving more and better housing. In the terminology used here, the subgoals are means or methods for achieving the mega goal, the sector's functional specialization.

2. The discovery that plastic pipes may be carcinogenic simply provides another stimulus to the debate and interactions within the sector.

3. Even though sectors operate separately, they can affect each other at the margins. For example, the parks-open space sector can serve as a watershed to protect water quality and protect areas for wildlife and biodiversity (environmental protection sector). Nonetheless, most government practice and policy reformulation treats sectors as distinct.

4. Kingdon (1984) and March and Olsen (1976) argue that technologies (solutions) search for sectors and problems as well.

5. The reasons that anachronistic or unresponsive governmental programs and agencies survive seem legion and intransigent. Aside from the interests of the people who administer the program, related special interests, and their clients, legislation itself (which may be challenged by sunset laws) perpetuates organizations. See Downs (1966) for a range of additional hypotheses. Chapter 6, "Delusions of Certainty and Their Consequences for Expectations of Government," offers another interpretation. Some programs may fade from public consciousness and power to mere ghosts of their former selves, however. Community action programs and model cities are good examples.

Sectoral Dynamics

6. To examine the hypothesis, consider the exceptions: when agencies within a sector compete, and when area governments collaborate. When agencies within a sector compete for resources and authorities within their specialization, their competition takes various forms. The competition appears to take place through understood rules of the game and actors realizing they will work together in the future.

Examples of area governments collaboration arise when neighboring jurisdictions deal with shared (common) resource pools, or the same specialization. For example, cities on the same lagoon collaborate on lagoon management, and police in neighboring jurisdictions cooperate on fighting crime on their shared borders. Area government collaboration is partial, focusing on solving a particular problem. Moreover, their geographical proximity ensures they will need to work together in the future. In each example of territorial government cooperation, the officials share both the same geographic resource problem and the same specialization (sector).

More complex cases develop when actors share a common resource pool and compete for external resources. Godschalk (1992) reports several jurisdictions sharing a limited watershed while some pursued aggressive economic development, which further strained the watershed. Similarly, the San Francisco Bay-Sacramento Delta pits environmental, industrial, and agricultural interests against each other in lengthy planning and negotiation. In the first example of shared and competing resources, planners and officials reached consensus. In the second example, stakeholders eventually jointly supported and succeeded in passing a $995 million bond measure for restoring and improving the Bay-Delta and for wastewater treatment, water supply and conservation, and local flood control and prevention.

Situations of shared resources or competition for external resources are not necessarily zero-sum games. For example, cities in a metropolis may compete among themselves for the location of a laboratory for a university that is already located in the metropolis, and collaborate to bring outside industries into the metropolitan area to expand the metropolis's pie.

Thomas (1994) reports the following still more complex case of intergovernmental collaboration in response to threat to agencies' turf:

> This information and knowledge [species survival depends on preserving sufficient portions of its habitat] combined with the regulatory threat of the Endangered Species Act to foreclose the possibility of certain actions within these jurisdictions led agency officials to believe that coordinated strategies can preserve more viable habitat overall (i.e., larger, more compact, and contiguous parcels) with less administrative and political costs to any one agency. (p. 15)

This case elaborates cooperative behavior with a shared resource pool and common threat trying to have greater long-run control.

7. The question of interjurisdictional equity is troubling. The perennial question of how distributional and allocation criteria can be reconciled becomes very important in intergovernmental fiscal affairs. Intergovernmental fiscal decisions often have intended and unintended consequences for interpersonal equity. At the same time, the frequent goal of interjurisdictional equity is controversial. Because, under conventional assumptions, interjurisdictional equity cannot achieve interpersonal equity, much normative literature tolerates the complexities of interjurisdictional redistribution on the grounds of political expediency. Criteria for measuring equity, whether interjurisdictional or interpersonal, are in dispute. What should be equalized: measures of per capita welfare, service standards, budget deficits according to fiscal effort and need, or some other criteria?

8. Yin's (1979) data support a related conclusion he does not reach explicitly: Counterpart bureaucracies respond to their sources of revenue.

9. Indeed, in today's globalized economy, some cities, such as Tokyo, New York, and Los Angeles, create placelessness while they retain their own primate status (Sassen, 1991). Also, place matters in terms of networks that create certain geographic regions and corporate cultures and dealings with their local governments (Saxenian, 1994). Moreover, local interorganizational networks develop among those dependent on community growth (e.g., local business, financial institutions, local media, realtors, and lawyers; Scott, 1992, p. 219).

10. Some would argue that area governments establish goals and then seek out the means to reach them, in a form of rational behavior. A case can be made instead that area governments proceed more cybernetically, moving toward sectoral benefits and away from sectoral harms, in reactive rather than proactive behavior, without any preformulated goal direction. This interpretation is not strictly necessary to the point that area governments do not determine sectoral effects on them. When local, areawide, or state governments initiate sectoral activity, they do so within their respective sector.

11. Again, this simpler question raises many questions about values and distribution.

12. Forester (1969) shows that changes in the present may have counterintuitive results when projected into the future. Of course, the consequences of the errors could become less important over time.

CHAPTER 6

DELUSIONS OF CERTAINTY AND THEIR CONSEQUENCES FOR EXPECTATIONS OF GOVERNMENT

This interpretation of U.S. intergovernmental structure and dynamics shows the institutionalization of sectors' technological trials to be costly, harmful, and restrictive of choice. At the same time, these trials and errors seem to serve a latent learning function—if at the expense of proliferating government and enduring effects on places and people.

This chapter speculates about why the intergovernmental system persists in this perverse form of learning. What prompts and perpetuates the institutionalization of technical trials? How do intergovernmental dynamics account for the system's failure to meet popular expectations of government? This chapter argues that these patterns are partly caused by the incompatibility between government's predispositions to certainty and the essential uncertainty characterizing its tasks.

Some would argue that there is (or ought to be) no contradiction between governmental needs for certainty and its tasks: Government undertakes (or ought to undertake) only those activities of which it is sure. In this view, like any private business, government must know its purpose and how to achieve it.

But the public business is fundamentally unlike private business, and entails much uncertainty. Government organizations originate in response to problems; if they fail to solve them, the organizations nonetheless continue to try.[1] Moreover, another government job is to correct market failures (e.g., uncertain externalities) and take responsibility for precisely those activities for which private enterprise's production functions fail (e.g., passenger rail service). Furthermore, government must provide unprofitable services, such as roads, which are necessary to maintain business itself, and thereby taxes to support government. Even more fundamental, government must deal with such matters as equality, justice, and future generations, which are entirely beyond business's domain. Thus, government must do what business does not know how to do. Government's work is full of uncertainty.

This chapter begins by examining how government's problem conditions vary in uncertainty and how expectations of governmental performance ought to correspond to these varying conditions. It then explains government's predispositions toward certainty. Next, it analyzes the intergovernmental system's dynamics in terms of the contradiction between government's actual conditions of uncertainty and its predispositions toward certainty. The chapter concludes that delusions of certainty propel the unfortunate institutionalization of sectors' trials and errors. These delusions permit sectors to pass the burden of actual uncertainty onto real places and people.[2]

VARIABLE PROBLEM CONDITIONS WITH RESPECT TO EXPECTATIONS FOR GOVERNMENT

Problem Conditions and Associated Expectations

This section sets out a matrix (Thompson & Tuden, 1959) adapted to clarify variations in public problem conditions (see Figure I.1). At the outset, it is important to note that the matrix's key variables—means, ends, and uncertainty—go to the heart of public decision making. The matrix is divided along two dimensions. The vertical dimension is technology, meant very broadly as the knowledge of how to do something, instrumental knowledge, or means. The horizontal dimension is goal, the purpose, desired outcome, or end. Each is dichotomized according to certainty-uncertainty. A technology can be known or unknown; that is, means have proven effective for a particular goal or they have not. This dichotomy thus corresponds to mature and immature sectors. Because a goal is value laden and thus cannot be proven known or unknown, goals are divided into agreed or not agreed.[3]

The matrix produces four prototype problem conditions: (1) known technology, agreed goal; (2) unknown technology, agreed goal; (3) known technology, not agreed goal; and (4) unknown technology, not agreed goal. Each set of problems is discussed briefly to clarify the conditions and to identify appropriate expectations and responses.

Known Technology, Agreed Goal

Known technology and agreed goal constitute the presumed conditions for classic bureaucracy (Gerth & Mills, 1946, pp. 196-244). When both means and ends are certain, a public action may be programmed. The particular set of procedures will then be repeatedly applied. For example, once the Salk polio vaccine was invented, it was given in prescribed doses to millions of children, rich and poor, everywhere. Soon after, the dreaded nationwide epidemic was eradicated. With such certainty, results are predictable. By treating like situations in the same way, a program can be equitable. Because the reliable technology can ensure the same outcome, it can be not only replicable but also accountable, efficient, and effective.

Under these conditions, the public product or service is achieved through repetitive administration according to prescribed formulas and procedures. Although some sectors have known means, particular applications depend on particular environments. A road, for example, must be tailored to topography. In such complex situations, decision authority is appropriately decentralized. Still, the basic technology is known and, accordingly, can be effective.

It is easy to take conditions of an agreed goal and known technology for granted. Examples abound but go unnoticed precisely because they are effective. Our capability to provide sanitary water and sewers meets our desire for these and numerous other collective goods and services.

Conditions of an agreed goal and an effective technology cannot be relied on to last forever, however. Consider the plight of Social Security, not so long ago a fine example of known technology (withholding tax and check writing) and agreed goal (retired workers should be cared for in their old age). But demographics changed. Now there are insufficient current workers to support the growing legion of elderly. As the technology collapsed, strong agreement on the goal persisted. A swift bipartisan effort restored Social Security. Yet the demographic shift ahead (as baby boomers retire) poses a major threat to Social Security.

Public opinion on the appropriate means to solve a problem may also change. Methods of mitigating poverty, for example, have shifted from the poorhouse approach to check writing, which in its turn was temporarily supplemented by a bold but uncertain strategy to build poor people's power, and is currently being transformed into decentralized "workfare."

Even when theory seems well grounded and effective, shifts in social values may threaten continued operations. For example, publicly supported[4] savings and loan institutions may have agreed on their objective of safe investments with good returns and known exactly the loan criteria necessary for achieving their objective. Relying on actuarial tables and avoiding risky neighborhoods, they made routine loans. But public opposition to their discriminatory method of "redlining"[5] minority neighborhoods threw previously stable organizations into turmoil. Today, they provide grants for affordable housing in neighborhoods they earlier bypassed to comply with the Community Reinvestment Act. Thus, stable conditions of certainty can disintegrate and then reform.

Although the conditions of agreed goals and known technology cannot be expected to be permanent, they do permit dependable public administration for some time and offer the necessary conditions for bureaucracy. This prototype situation and model organization allow governments to meet popular expectations for predictability, equity, accountability, efficiency, and effectiveness.

Unknown Technology, Agreed Goal

The problem of high African American teenage unemployment typifies the genuine concern and seeming intractability of public commitment to address pressing problems (an agreed goal) without a proven solution (unknown technology). The agreed goal is reducing African American teenage unemployment. The immediate problem, however, is that as yet there is no proven method for achieving the goal. (People may propose plausible approaches, but these approaches are far from having a tried-and-true effective track record.) Thus, in contrast to the first set of conditions, where knowledge is already in hand, this situation calls for attaining the missing knowledge.

Many public problems fall into this category of agreed goal and unknown solution. Examples range from finding a cure for cancer to reliably teaching disadvantaged children to read, from safely disposing of toxic wastes to integrating ex-convicts into society.

Uncertainty over means must be approached through experimentation. The search for effective means may be explicit—perhaps through research and demonstration or through pure research. For example, government funded a controlled, expensive, nearly decade-long experiment (RAND Corporation, 1982)[6] to learn the effects of housing allowances on housing quality and housing markets.

More often, the search is pragmatic. Thus, experimentation in public programs customarily aims not at developing causal theory but rather at finding an instrumental solution, something that works. Usually a sector tries something slightly different than usual and waits to see what happens. If not scientific, at least this process seems expedient. But unfortunately, each new trial is set in institutions and widely applied as if it had already been proven effective.

Conditions of known problems and unknown solutions demand innovation to encourage learning. For example, when a sense of urgency provided political support, government amply funded many experiments on alternative sources of energy. Such conditions call for inventiveness and creative sensitivity to varying constraints. Innovative responses yield information that contributes to the development of an effective technology.

This situation is entirely different from that in which the technology has already been proven effective and, accordingly, the problem has already been solved. Consequently, expectations of government performance ought to be different as well. By definition, innovation is novel and therefore contradicts routine predictability. Accordingly, the public cannot demand accountability for results. If an innovative program for mentoring welfare recipients fails to result in job placements, for example, its inventor is not to be punished, but rather rewarded for providing new information.

Placing a premium on sensitivity to difference precludes equality; different constellations of people and places will necessarily elicit different innovations. A program for disadvantaged toddlers in Marietta, Georgia, for example, would provide different experiences than one in the South Bronx.

Finally, if a method is unpredictable, management for efficiency and effectiveness is meaningless. Thus, officials addressing an unsolved but pressing problem ought to be held accountable not to standard bureaucratic norms suited to solutions that are already programmed. Instead, officials should be expected to conduct an open, innovative search for a solution.

Known Technology, No Agreed Goal

For a simple example of effective proven methods in the context of disagreement over goals, an economic development expert knows how to develop a particular site for labor-intensive industry, and an environmental group knows how to preserve the ecology of the same site; they merely disagree on the goal. Or the energy sector knows how to establish a liquefied natural gas terminal, but the citizens of the selected community resist it. Such conditions represent differences in preferences.

The appropriate response to these conditions is bargaining. Differences in process and political influences can make the inherent conflict more striking or more submerged. Thus, bargaining may mute and obscure potential difficulties, or it may compensate various interests through trades, or it may adapt technology to achieve several goals at the same time. Regardless of the particular form and effect of the bargaining process, the expectation is clear: to accommodate multiple preferences.

Each completed bargain must be tailored to its particular participants, their issues, their context, and their preferences. Because each deal is thus unique, the bargaining process is antithetical to bureaucratic routines yielding identical results. Adjustments necessary for bargaining fly in the face of bureaucratic norms dependent on inviolable rules. Participants in negotiations cannot and ought not to be held accountable for predictable, uniform outcomes. Instead, they should be expected to accommodate diverse preferences.

Delusions and Their Consequences

Unknown Technology, No Agreed Goal

Conditions resulting from multiple, or unarticulated, preferences and no effective means for meeting them are troubling, for they can disintegrate into anarchy. Consider the urban riots of the mid- and late 1960s. Ghetto residents felt things were terribly wrong, but they had no way to set them right.

Even in less extreme cases, these conditions of uncertainty are disturbing. Clear examples are impossible because these conditions are in chaos. Nevertheless, such conditions are common. Goals are often nebulous and changing; facts are often ambiguous. For example, the San Francisco estuary project entailed considerable uncertainty over both means and ends. Many interests—including multiple government agencies, developers, environmentalists, and scientists—were involved and mistrusted one another, worrying about what the others would do to a valued, shared resource. The nature of the resource itself was ambiguous (Gruber, 1994). Collective interpretation of this uncertainty is sometimes vaguely discomfiting, in a sort of unarticulated alienation, sometimes confounding.

The appropriate response to these "wicked" (Rittel & Webber, 1973) problem conditions is to facilitate social learning. Social learning can emerge from transactive planning (Friedmann, 1973, 1987), negotiation (Susskind & Cruickshank, 1987), interactive planning (Forester, 1989; Innes, 1995), and other kinds of participatory planning practice. Clarifying goals or problems or focusing on workable technologies through social learning shifts the problem to more manageable conditions. For example, despite high levels of mistrust, facilitated social learning in the San Francisco estuary project eventually led to agreement on an indicator of estuary health (salinity), thereby clarifying the previously ambiguous nature of the resource. Of course, social learning can break down or become mired in dead ends, but when it works, it shapes the chaos into some order or at least sets some direction.

Conclusions Regarding Variable Problem Conditions

Figure 6.1 summarizes variable problem conditions, the prototype process for treating each set of conditions, and corresponding expectations of governmental performance.

Figure 6.1 clarifies several issues relevant to the intergovernmental system. First, three of the four sets of public problem conditions

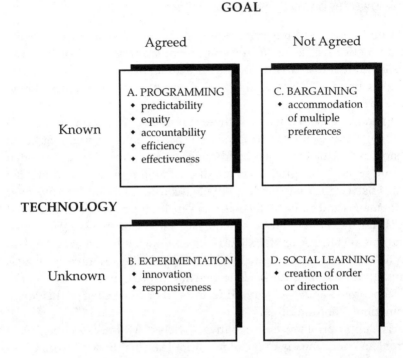

Figure 6.1. Expectations of Government Associated With Prototype Conditions of and Responses to Public Problems

explicitly address uncertainty—over goals, technology, or both. Even public problems with a certain goal and proven technology are vulnerable to eventual uncertainty. Consensus behind either the goal or its effective technology may erode, and a once certain, stable administration may be subject to environmental turbulence. In short, uncertainty characterizes all governmental problems.

Second, each set of conditions calls for a different set of responses. Although such a conclusion seems obvious, it is nonetheless alien to the practice and expectations of government. Governmental activities tend to be programmed (the theoretically appropriate response to certainty) regardless of actual conditions. Practices appropriate to conditions of uncertainty seem rare and somewhat questionable. Although negotiation is gaining credibility in practice (Godschalk, 1992; Hoch, 1994;

Innes, 1992) and is recognized in the literature (Bingham, 1986; Susskind & Cruikshank, 1987), negotiation remains dubious in the public eye. Despite nationally endowed research and development, explicit experiments on public problems seem even more suspect.

Third, each problem response calls for a different expectation. Moreover, some expectations are mutually exclusive. For example, expectations appropriate to programming, such as uniform predictability, are inappropriate to bargaining; prescribed outcomes would mean that nothing was negotiable. Or people might demand both effectiveness and innovation. Thus, for example, a governor tries to close an energy agency on the grounds that it failed to produce oil, when it had been operating under a different mandate: to devise new sources of renewable energy. Because the appropriateness of particular expectations of government depends on particular conditions, all expectations cannot be aggregated and espoused simultaneously.

GOVERNMENTAL PREDISPOSITIONS TOWARD CERTAINTY

In itself, the preceding point accounts for some of the discrepancies between expectations of government and its actual performance. Expectations of governmental performance ought to correspond to problem conditions. For example, government cannot and ought not to be held accountable for efficient predictability and innovation in the same program at the same time. To complete the case of the energy agency, the governor might do well to add a separate unit for certain production and let the uncertain research continue. A single agency may address different problem conditions. Some of its actions are routine and others are not.

Popular and political demands for contradictory expectations of governmental performance persist. These demands are understandable given public hopes—or illusions—that (1) government knows what it is doing (known technology), and (2) what it is doing is good for everybody (common goal).[7] Such mistaken presumptions of certainty reinforce systemic forces, which disguise actual problem conditions.

This section explores administrative, legal, and political forces that predispose government toward certainty regardless of actual problem conditions. It describes typical behavior arising from delusions of certainty: premature programming and premature consensus.

Administrative

Four aspects of classical administration suit it to presumptions of certainty. First, its definition, "dispensing or tendering to another . . . according to a prescribed formula" (Webster-Merriam, 1956, p. 12) implies a certain production function. This model of administration is bureaucracy (Gerth & Mills, 1946, pp. 196-244). In principle, bureaucracy treats every like case the same and so yields not only predictability but also equality, a kind of procedural justice.

This same predictability endows bureaucracy with its productivity. Tasks are rationalized, subdivided, and assembly lined. Bureaucracy supplies an organizational machine for doing the government's work. It depends on rationality and prescribed rules to achieve predictability, equality, accountability, efficiency, and effectiveness. Bureaucracy works through certainty.

Second, however, an organization is "an open system, indeterminate and faced with uncertainty, but subject to criteria of rationality and hence needing certainty" (Thompson, 1967, p. 10). An agency's actual circumstances are nebulous and shifting in complex intergovernmental relations. But expectations of predictability and productivity contradict this reality and demand dependable order. Organizations' basic production functions "rest on closed systems of logic" (Thompson, 1967, p. 23) that must be buffered from external influences. Thus, "paradoxically the administrative process must reduce uncertainty" (Thompson, 1967, p. 158).

Third, administration concentrates on controlling. Control means "the ability to determine a class of events or a state of affairs. . . . The extent of our ability to control is, accordingly, a function of our knowledge" (Landau & Stout, 1979, p. 149).

These three aspects of administration—bureaucracy's dependence on certainty, its actual uncertain environment, and its consequent emphasis on control—are tightly bound. Organizations need certainty to be themselves, to fit their definition and their model. They must act by prescribed rules to fabricate a sort of certainty and then exert controls to ensure adherence.

The point here is simply that because administration is compelled to reduce uncertainty, it mistakenly concentrates on control systems even though they "provide protection against infractions of rules—not against error" (Landau & Stout, 1979, p. 153)—and thus impede rather than encourage problem solving. This aspect of administrative bias toward control as a pseudo-substitute for certainty underscores the meaning of control, "to exercise directing, guiding, or restraining power over" (Webster-Merriam, 1956, p. 181). However misguided, administrative needs for certainty and control stem from demands for accountability.

Fourth, professional expertise predisposes administration toward certainty in another way. Professionals are closely tied to most sectors and carry authority because of their expertise (Parsons, 1960; Perrow, 1970; Thompson, 1967). To enhance this power and stature, then, it behooves professionals to present their expertise as if it were certain. Moreover, professional training and associations tend to reinforce professionals' own belief in the certainty of their expert knowledge.

In sum, the bureaucratic model, the closed logic of production functions, the drive for control to both reduce uncertainty and ensure accountability, and professionals' conviction of their expertise all bias administration toward certainty.

Legal

Legal predispositions toward certainty stem from two connected rationales. First, adjudication needs known laws against which behavior can be judged either legal or illegal. The legal basis of government depends on the certainty of "permanent" rules for its long-term stability. People then have a reliable basis on which to act and to anticipate others' actions. Statutes and regulations may be vague or ambiguous in practice, however. Law need not be immutable; it changes and is reinterpreted constantly. Rather, at every particular moment, there must be a body of law against which individual and agency actions can be judged and justice can be achieved.

Second, this legal need for certainty offers an avenue for institutional reform. If the law clearly says a particular agency must perform a specified task in a prescribed manner, governmental actions—or inactions—may be challenged in the courts. Thus, suits alleging that environmental impact assessment procedures have not been followed can delay and ultimately prevent major projects. Indeed, environmentalists may prefer specific laws that leave implementing agencies scant discretion (Adler, 1990). Legal recourse thus offers a check against administrative abuse.

Like administrative efforts to control, however, legal efforts can backfire by constraining variety, choice, and learning. In this sense, scrupulous legal specificity constitutes another delusion of certainty. In design, for example, environmental impact statements were meant to be open explorations, attempts at anticipating future consequences of proposed actions. Yet repeated legal challenges and decisions have tended to prescribe, specify, and close the environmental impact procedure. Regulatory guidelines became recipes and court cases became bases that must be touched.

Political

Political predispositions to certainty, unlike administrative and legal biases, are rooted less in a need for operational or formal certainty than in a need for symbolic certainty to inspire confidence. Politicians act in ambiguities rather than in the technological algorithms sought by administration or the clear-cut right or wrongs sought by judicable legislation.

Nonetheless, norms of political behavior presume certainty. Politicians tend to propose solutions[8] to the problems they are quick to identify; they cannot build a platform or voter enthusiasm from admissions of ignorance. Imagine the gravest uncertainties and the best politician. Franklin D. Roosevelt all but erased these uncertainties by pronouncing that "we have nothing to fear but fear itself." The rule orientation of legislators and administrators reinforces this political propensity to avoid articulating uncertainty.

Behavior Arising From Delusions of Certainty

The contradiction between actual public problems' characteristic uncertainty and administrative, legal, and political predispositions toward certainty is not resolved in the U.S. intergovernmental system. Instead, the contradiction is buried in delusions of certainty. Actors in all parts of the system tend to behave as if (almost) everything is certain. Moreover, the institutional structure in which they act induces an artificial, symbolic, and legitimated certainty. Powerful specialized sec-

Delusions and Their Consequences

tors define, subdivide, and isolate problems in such a way that they become cloaked in illusions of certainty.

The process of formulating such pseudo-certainty can be posed in terms of the problem matrix. On the technological dimension, the system is prone to premature programming (Landau, 1973; Landau & Stout, 1979). That is, from a narrow set of alternatives, a particular technology is selected and then implemented through programs, rules, procedures, organizational mandates, and so on, as if the technology were certain and proven effective. By being cast into Box A in Figure 6.1, it is prematurely programmed. On the goal dimension, the system is prone to premature consensus. That is, the fragmentation and specialization of the U.S. institutional context skews democratic access and deters debate over goals in such a way that each sector's goal appears acceptable. By being cast into Box A, it has arrived at premature consensus.

Policies are adopted and carried out as if solutions were known. Thus, these policies take the form of rules and procedures to be faithfully executed by sets of interlocking bureaucracies. Moreover, because the program is understood as the solution, the rules and agency relations are cast in formal laws and regulations and treated as permanent and inviolable. Yet the rules derive not from proven, effective, mature technologies but rather from untested, usually plausible hypotheses. Therefore, this premature programming inappropriately aims at unachievable standards such as predictability and accountability.

Such standards, their consonance with the bureaucratic model, and administrative, legal, and political predispositions toward certainty induce some public legislators and administrators to concentrate on rules.

Efforts to impose rules to create good policy seem to follow two lines of faulty reasoning. First, some begin not with the policy but with its traditional vehicle, bureaucracy. Recognizing that bureaucracy relies on rules, decision makers conclude that if they assign rules (Lowi, lecture at the University of California, Berkeley, May 15, 1978), they will achieve bureaucracy's promise of effective, accountable results. This kind of logical confusion corresponds to premature programming.

A second kind of faulty reasoning begins not with the policy but with larger values, equality of treatment. Recognizing that fairness demands inviolable rules, decision makers conclude that if they assign rules, predictable, equal treatment is bound to result.

Both lines of faulty reasoning reverse the logical order by presuming conditions of certainty. When a goal is agreed on and its technology is proven effective, it is suitable to being programmed into a bureaucracy. Such a technology contains its own rules, which allow it to operate effectively and fairly. When technology is uncertain, no imposition of rules can make a policy predictable or equal; its guiding rules must be discovered or invented.

The Personal Responsibility and Work Opportunity Reconciliation Act of 1996 (hereafter, called welfare reform) illustrates premature programming. In the 1996 election year, pressure to produce welfare reform was irresistible. The legislation depended on rules—for example, years on welfare and percentage of cases eligible for emergency extension. But the rules had no foundation in research, demonstration programs, or proven methods. The rules emerged from politics and became a premature program, sure to produce unanticipated results and further reformulations after harming real people.

Box 6.1

Community development history illustrates premature consensus. When urban renewal was proposed to ameliorate postwar city deterioration and white flight to the suburbs, it was supported widely. Not only chambers of commerce but also city officials looking for a stronger tax base, builders and developers looking for good sites, and even welfare advocates looking for improved living conditions all saw the proposal as beneficial. Nevertheless, consensus for urban renewal was premature. After the program had begun, after scores of neighborhoods and hundreds of thousands of housing units were destroyed, residents began to protest, and the program began to change.

Program changes came in the form of new rules issued to thousands of U.S. Department of Housing and Urban Development field staff in loose-leaf binders. Every time a problem necessitated an adjustment, it took the form of a new rule applicable nationwide. New pages of rules replaced and amplified old ones—the paper manifestation of repeated premature programming.

In sum, predispositions toward certainty overcome the reality of public problems' uncertainty and result in premature programming and premature consensus. Such false certainty is expressed in rules and procedures that are institutionalized as if they were derived from a proven technology. But because instead these rules and procedures represent only untested hypotheses, their across-the-board application results not in the predictable, equitable outcomes expected, but in unanticipated and often undesired effects. Indeed, it seems that sectors may be fabricating a dubious security of certainty for themselves at the expense of disadvantaged places and people.

INTERGOVERNMENTAL SYSTEM DYNAMICS AND DELUSIONS OF CERTAINTY

This section explores how this process of shifting uncertainty occurs through the sectors' premature programming. The intergovernmental system is assessed in terms of expectations of governmental performance.

Programming and Stable Sectors

When a sector's technology is mature, interactions among its institutions are stable and ordered. Routine tasks subdivided among the institutions resemble those subdivided among units within a bureaucracy. To the extent that a sector's technology is effective and the general public continues to support the sector's goal, the entire set of agencies and their interactions can be programmed.

The bureaucratic model of programmed intergovernmental responsibilities need not be centralized. Social Security, for example, has centralized rules but operates through field offices to ensure easy access for clients. Other stable technologies—for example, road building—must be decentralized to adapt to different local conditions.

Stable sectors have mature technologies and supported goals. They depend on rules to replicate their products and services in multiple situations. Reliance on effective rules permits stable sectors to operate according to a bureaucratic model. Their proven technologies enable them to meet expectations of predictability, equity, accountability, efficiency, and effectiveness.

Despite denigrations of bureaucracy, this situation is desirable because a mature sector provides a wanted governmental good or service reliably. Such sectors present no problems and thus demand no public attention or debate. Their dependability makes them almost invisible.

Only when the public begins to question a technology or goal do such stable sectors encounter uncertainty. This point suggests that sectors' salience varies over time. Recently, for example, public utilities, which once were so stable as to be scarcely noticed, have come under public scrutiny because of crises over energy and conflicts over the multiple means of producing it. Although maintaining their general sector mission, they shifted the form of their goal from increasing the supply to reducing demand through conservation.

In short, programmed relations among governmental units in stable sectors serve public interests well, but only so long as their effective technologies and goals retain public support.

Experimentation and Sectoral Interaction

Many sectors have uncertain, or ineffective, technologies. Their operations entail frequent surprises, crises, and criticism, which stimulate chains of agency interactions. The resulting revised and expanded rules and organizations constitute an institutionalized search for an effective technology.

Although the circumstances are common, only rarely can a sector face—or articulate—these conditions of a pressing public problem with no known solution. These conditions call for experimental decision making, with corresponding expectations of responsiveness and innovation. Moreover, they defy rules and call for antibureaucratic organizations. The Community Action and Model Cities programs of the affluent, committed, mid-1960s exemplify the unusual situation of pressing problems, coupled with the public recognition that they were unsolved. In their early years, these programs were granted the flexible structures and freedom from rules they needed to foster responsiveness and innovation.

Usually, the intergovernmental system cannot confront the demands posed by these troubling conditions because political, legal, and administrative behavioral norms presume certainty. Moreover, such situations challenge the core of sectors' dominance, their supposed expertise through specialization.

Public issues rarely present themselves de novo; they usually stem from a sector's preexisting rules and interactions. It is specialized civil servants, not voters, who tend to initiate additions to the public sector (Beer, 1976, pp. 160-162). Thus, a de facto search for an effective technology is cast against the backdrop of a bureaucratic structure resting on a presumably advanced, or at least experienced, technological base.

Partly for this reason, and partly because of vested interests' hold on the sector, the search for a new solution is liable to be formulated as a marginal adjustment to standard operating procedures. The agencies tend to resist open, expansive experiment as disruptive to their fairly smooth, or at least familiar, interorganizational political and personal relations.

The effect of such a constrained search is that the ideas that emerge are only modest adjustments to the preceding, ineffective technology. The cautious new versions are then prematurely programmed and continue to reinforce the orientation and preferences of the professionals and special interests aligned with the sector. Consequently, when a problem seems intractable, some potential—but again untested—solutions are politically infeasible. The choice of which potentially effective technology to adopt is constrained by the powerful networks supporting each sector. At bottom, the choice is often disguised as technical and specialized, but it is prestructured and thus in some sense political.

In employment, for example, multitiered minimum wages is an impossible policy, but a limited summer jobs program at the uniform, high, minimum wage is possible. (Granted, multiple minimum wages is a contradiction in terms, but the idea of allowing lower wages for teenagers and alien workers has been proposed.) Through early battles and a long history, organized labor has built the labor sector. Unions, the Department of Labor, and their interpretations of workers' best interests have set constraints on the kinds of jobs programs and regulations that can be adopted.

Presumptions of certainty and historically embedded specialization together act to program narrowly construed policy. The resulting rule-bound agencies typically limit modest improvements to minor variations within the sector. In their turn, these new versions of the sector's programs and institutions dampen the prospects for an open, experimental search for effective technologies.

In sum, forces in the intergovernmental system act against and disguise the need for innovation. When pressing problems without solutions exist, adjustments to public interventions are usually prematurely programmed. Applying unproven technologies gets surprising results. They trigger interactions that constitute an unintended, institutionally constrained search for an effective technology. Thus, the drive for certainty binds the prospects for change to established agencies and their powerful supporters, and so limits the very innovation essential to resolving the uncertainty. This systemic contradiction results in costly, distorted, and unresponsive programs.

Bargaining and Sectoral Effect on Area Interests

As discussed in Chapter 5, "Sectoral Dynamics: Institutionalized Technological Elaboration and Effects on Area Interests," sectoral effects on areas are disjointed. For political and administrative reasons, area interests are usually perceived and articulated along sectoral lines. Area interests are a jumble of concerns that are formulated and reformulated with respect to each particular sectoral imposition.

Overall, segmented responses to sectoral demands suits pluralism. In this view, pluralism means that no single (elite) or unifying (homogeneous) interest prevails, but rather that different interests and coalitions of interests assert themselves at different times over different matters (Dahl, 1961; Truman, 1951). Because preferences vary among constituents and over time, different clusters form different interpretations of area interests in response to disjointed sectoral demands. Thus, actual intergovernmental dynamics seem to meet appropriate expectations for accommodating multiple preferences.

Behind this simple congruence, however, lies an important complication. Even a single sectoral demand on an area is apt to incur some resistance, because it affects diverse preferences. In this light, sectoral demands on area interests correspond with the public problem conditions of a known technology and no agreed goal. Because (barring a homogeneous, stagnant community) differences are all but inevitable, bargaining or consensus building is critical to the juncture of sectoral and area interests.

The form of intergovernmental exchange can vary substantially; interaction may look like collaboration or contention. In the end, however, the intergovernmental relations play themselves out into some form of resolution. If the exchanges have been particularly divisive, the eventual agreement may be an arbitrated settlement, some quid pro quo deal to avoid stalemate. For example, a town will accept a minimum security women's prison, but only on the condition that it be built to blend in with local housing. In some special cases, more collaborative intergovernmental efforts may result in new, mutually enhancing arrangements, for example, specially tailored parks and transit.

In this sense, actual intergovernmental system dynamics are compatible with the expectation of accommodating multiple preferences. In fact, when a mature technology adapts to area conditions and preferences, it combines its sectoral knowledge with the area's knowledge of the particular place to achieve joint goals by both using and responding to diversity.

The seeming compatibility between actual sectoral-area interactions and the accommodation of multiple preferences masks two important systemic biases, however. First, confrontation between sectoral demands and multiple area preferences favors the sector. Area response is customarily cast in terms of narrow, sector-related perspectives. Political and administrative propensities tend to reinforce the segmented approach, which serves sectoral interests. Peterson, Rabe, and Wong (1986) find this pattern of sectoral dominance to be more pervasive in more professional sectors, such as medical, than in less professional sectors, such as housing. Even in housing, however, local political concerns play out through the housing program administration, that is, within the housing sector. In effect, Peterson et al.'s research suggests that it is easier for cities to resist or adapt impositions of sectors that are less professional.

Other causes of sector dominance in sector-area bargaining stem from skewed representation. Although the sector is integrated, area interests are multiple, diffuse, and sometimes structurally at odds with one another. Even when substantial losses or gains are at stake and multiple interests can be coalesced, ad hoc groups tend to be disadvantaged with respect to a long-established sector. Diverse local interests must first discover common bonds and organize, then strategize, then learn a technology before they can debate effectively with a strong, institutionalized, well-researched, and monied sector. Moreover, sectors are powerfully connected through their institutionalized ties to elite special interest groups and legislative committees.

As sectors initiate interaction, they formulate the problem from their expert information and structure the bargaining situation. Potential area opposition, formed from a loose coalition of multiple, partly diffuse interests, can rarely afford to propose an explicit counter plan for fear of fractionating the coalition's diverse membership. Instead, all they can do is protest, "No! Not in my backyard" (NIMBY). Occasionally they do stop a sector's imposition, for example, the San Francisco freeway revolt and failure to site an oil refinery on the East Coast (Popper, 1987, p. 8). More often, local protest tends to delay projects and force sectors to adjust projects to local concerns.

Finally, area networks with potential common concerns are often composed of multiple jurisdictions whose interests are severed by differential (sometimes discriminatory) residential, industry, and taxing practices. This structurally induced competition (Kenyon & Kincaid, 1991) inhibits joint area strategies in response to sectoral demands. Table 6.1 summarizes and contrasts characteristics of sectoral interests and area interests.

For all these reasons, the sector usually holds the upper hand in sectoral-area bargaining. Consequently, intergovernmental dynamics distort accommodation of multiple preferences by biasing the process toward sectoral interests.

The second systematic bias seems to overwhelm earlier conclusions. Although sectoral-area dynamics permit some adjustments to competing preferences, the actual process systematically avoids accommodating multiple goals. Because each sectoral imposition becomes a segmented bargain, biased toward the sector, nearly every accommodation is no more than an adjustment of technology. The sector's good or service is modified—more or less—to suit an area's conditions or preferences. For example, a job training program may adjust to a booming area's demand for highly skilled office workers. Technological adjustment is usually modest, and the sector, perhaps learning something along the way, prevails. Thus, the dynamics of the intergovernmental

Delusions and Their Consequences

TABLE 6.1 Contrasting Characteristics of a Prototype Sector's Interests and an Area's Interests

Sector's Interests	Area's Interests
Single, integrating goal	Multiple, diffuse, sometimes competing goals
Long-established	Ad hoc coalescing
Training and socialization reinforce integration	Vulnerable to disintegration
Experience with its technology	Must learn sector's technology
Established sources of power and money	Must invent or attract sources of power and money
Initiate—therefore formulate on own terms	React to sector's formulation

system direct debate to the means of achieving the sector's goals and virtually preclude trade-offs between sectors and thus bargaining over ends.

Confusion Stemming From Prematurely Programmed Sectoral Effects on Diverse Area Conditions and Preferences

When a sector's technology is immature, it is engaged in testing, revising, and retesting. Therefore, the sector cannot control the consequences of its actions. Because it can neither predict the effects its immature technology might impose on area conditions nor adjust this technology to area preferences, such a sector is ill-suited for bargaining.

Applying an unknown technology where no goal has been agreed on typically results in confusion. Fortunately, because the intergovernmental system contains many parts that work well through programming and bargaining, this partial disarray does not disintegrate the entire system. Its immense variability and complexity absorb some confusion. Nevertheless, because immature technologies frequently confront multiple area preferences, a good portion of the intergovernmental system seems in chaos.

The most obvious form of uncertainty over both means and ends in the intergovernmental system occurs when a sector applies an unknown technology broadly without agreement on its goal. This process results from delusions of certainty. Problem conditions of uncertainty over both means and ends arise and take expression in the intergovernmental system in three additional ways. These processes vary in their degree of problem articulation and thus of seeming chaos.

The least obvious form of uncertainty over both means and ends in the intergovernmental system occurs when one tries to specify what is inchoate and unlabeled. This form of uncertainty exists because public policy conditions are ambiguous. Some issues are attended to, articulated, categorized, and treated. Others are neglected or fall through the cracks of other problem formulations, and so are left nameless and thus uncategorizable and untreatable (Douglas, 1986). A sector might cast its technology in a way that eludes latent facets of its problem.

For an oversimple example, until fairly recently women's work in the home had no public policy label. It did not count as "work." When women worked outside the home, their work fit the traditional label. Meanwhile, needing help in the home, these women hired workers (often women) who worked outside their homes performing erstwhile nameless services. This growth in services—cleaning, child care, cooking, and much more—and specialization of work that had previously been uncategorized has confused gross national product accounting and labor policy. The case illustrates how an issue can fall through the cracks of other problem formulations and how a policy can be framed to elude latent facets of collective concern.

Another less amorphous form of uncertainty over means and ends may arise when conditions are perceived as rife with problems that for a period defy articulation. Thus, both goal and technology are uncertain. People may talk and despair, whether in a single neighborhood, in urban slums, or on national television. Through this interactive, social construction of reality (Berger & Luckmann, 1966), they try to make sense of the disturbing chaos; they try to identify and label actual problem conditions.

In another form of uncertainty over means and ends more attuned to the discussion here, groups with different goals and associated but immature technologies may be pitted against each other. Uncertainty over goals stems from multiple competing goals forced into trade-offs. Each group may be quite convinced about its own goal; the problem is simply that the groups disagree. The immaturity of each faction's technology confounds matters. It may be irrelevant how much of each group's technology is undertaken because none is predictable.

Delusions and Their Consequences

Because the intergovernmental system usually avoids pitting sectors—whether stable or unstable—against each other, this third form of uncertainty over means and ends rarely arises. An example is severe budget cuts in several social services.[9] Major funding loss can force cross-sector choices; which activities are less valued and can be abandoned so that others can be maintained? The decision is particularly difficult because the programs, such as drug abuse counseling, have immature technologies.

A more articulated and familiar form of uncertainty over both means and ends results from a prematurely programmed sector applying its trial technology on real places and people with diverse goals. This troubling situation would be much less pervasive without government's tendency to treat unknown technologies as if they were proven effective.

Premature programming generates a false facade of certainty. This presumption of certainty and programmed procedures inclines unstable sectors to act as if their technologies were indeed stable. Accordingly, they follow their rules, apply their programs across the board, and try to adapt them to area interests. But rather than achieving intended results, applying an unknown technology leads to new, surprising, consequences. These events stimulate interactions among the institutions and another set of rules and procedures.

At the same time, applying an uncertain but programmed technology to multiple diverse area conditions and preferences results in different formulations of the sectoral goal. Although the overall goal remains fixed (e.g., less crime), adjustments to varying area preferences may significantly alter strategies. Precisely because their technology is immature, preprogrammed sectors are likely to experience goal adjustments as well as surprising consequences.

(In contrast, a stable sector with a mature technology is immune from goal adjustments; its adaptation to area preferences, as well as its technology, is predictable. In effect, area conditions and preferences are simply variables in its reliable production function.)

The composite effect of an unstable but prematurely programmed sector's bargaining with area interests yields a seemingly endless cycle, with neither goals nor technology to provide a point of reference. The process perpetually alters the sector. But each cycle cannot be guaranteed to respond to area conditions and preferences, because each version constitutes but a trial of a revised technology and a revised strategy. Although the process is disorderly, it may eventually resolve itself by disclosing some recurring preferred problem formulations or some procedures that are repeatedly effective in certain circumstances.

CONCLUSIONS

In sum, the intergovernmental system is compatible with governmental predispositions toward certainty only when the application of a sector's proven effective technology encounters no politically overwhelming opposition. In this situation, programming can achieve expectations of governmental predictability, equity, accountability, efficiency, and effectiveness. When a sector's effective technology is applied to areas with multiple preferences, the technology may adapt. In this situation, bargaining can achieve the expectation of accommodating those preferences.

On the other hand, when technology is uncertain, the intergovernmental system works perversely. Sectors presume consensus and prematurely program marginal adjustments acceptable to the sector's professional and vested interests. This narrowly constrained technological elaboration deters a full, open, experimental search for innovations to resolve the uncertainty. When such prematurely programmed sectors impose their trials across the board in diverse settings and try to adjust them to multiple area interests, they generate cycles of trials and reprogramming. These new variations incur their own rounds of surprises and hardships. Over time, they may reveal some reliable results or some repeatedly popular formulations to provide some eventual grounding in certainty.

Figure 6.2 brings together the actual performance of the intergovernmental system with corresponding expectations of government. It captures the system's failure to meet some expectations for its performance.

The chapter has shown how delusions of certainty prompt premature programming. Potential solutions are limited to those acceptable to a sector's professionals and vested interests. This syndrome not only impedes innovation, it also reinforces sectoral dominance and feeds a cycle of costly, often harmful, trials and errors.

Delusions and Their Consequences

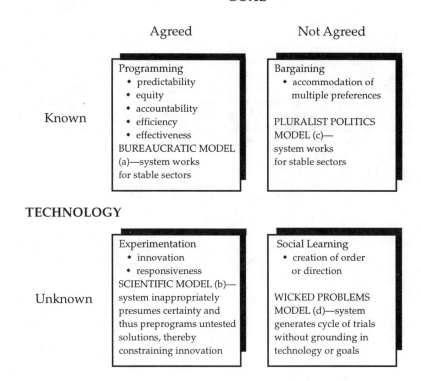

Figure 6.2. Summary of Intergovernmental System Performance With Respect to Variable Problem Conditions of Certainty and Expectations of Government
SOURCES: a: Gerth and Mills (1946); b: Popper (1959); c: Truman, 1951; d: Rittel and Webber, 1973.

NOTES

1. Public agencies may receive bigger budgets when they fail in their tasks. Conversely, public organizations that achieve their objectives are in danger of extinction and must diversify to new problems to survive.

2. Marris (1996) shows chains of actions progressively displacing uncertainty onto weaker organizations, neighborhoods, and people.

3. Of course, the real world is not this tidy. In theory and practice, the line dividing means and ends may blur. Furthermore, technologies are rarely precisely known or unknown, but rather over time show themselves to be more or less effective. Also, technologies may have varying levels of public support. Similarly, a goal may elicit various degrees of agreement. Even knowledge may be a matter of degree of beliefs.

4. Through Federal Housing Administration mortgage guarantees, among others.

5. Redlining is a practice of drawing a line on a map around certain areas where loans and mortgage insurance will not be granted because of risk assumed to be associated with conditions in those areas.

6. This work was complemented by studies on the demand side, mostly by ABT Associates. The experiment cost about $65 million.

7. Popular distrust of government, the recent "Contract with America" antigovernment presumptions, and bureaucratic inertia in the former Soviet Union and in many developing countries have popularized privatization and other entrepreneurial strategies. These coexist with government predispositions to certainty and programming.

8. Political behavior connected with elections, which polarize issues for the voting form of decision making, must be distinguished from ongoing political behavior connected with the negotiating function, which must obscure conflict for the bargaining and consensus forms of decision making. The effective politician reconciles these contradictory roles through very general campaign promises, but even so often reneges on articulated policies after election.

9. One might think that budgets would consistently pit one sector against another. But public decision processes have found ways to treat funding, and even funding cutbacks, incrementally (see Wildavsky, 1964).

CHAPTER 7

CONDUCTING PUBLIC POLICY IN CONDITIONS OF UNCERTAINTY

The preceding chapters examined the current structure and dynamics of the U.S. intergovernmental system. Specialization cuts across traditional territorial levels of government to forge sectors that link not only administrative levels but also associated legislative committees, nonprofit agencies, and vested interest groups. The forces propelling and elaborating specialization express themselves in bureaucratic institutions, in premature programming and premature consensus, and in sectoral dominance.

Increasing specialization reinforces delusions of certainty. "Specialization is the trained incapacity to think" (Veblen, quoted in Bardach, 1977, p. 127). The goal is determined by the definition of the specialty, and so effectively precludes uncertainty. For example, air quality experts know that the goal is to keep pollution below a specified number of particulates of specified types per cubic meter of air. Further specialization results in planning for particular polluting sources. As issues become compartmentalized, specialization creates a sort of pseudo-certainty.

This chapter proposes the strategy of variable, contingent planning processes and policies as a way to conduct public policy in conditions of uncertainty. Variability means the ability to vary, rather than having only one predetermined form. Planning processes and public policies can vary by taking different forms in different conditions and by generating different results. The strategy of variability draws on popular principles that defy bureaucratic standards. The chapter begins by showing how the strategy of variability undermines sectoral dominance and responds to uncertainty. The bulk of the chapter shows how to tailor planning processes, public policies, and organizations to problem conditions. The chapter concludes by tracing how variable planning and policies contribute to a more responsive policy-making system.

A STRATEGY OF VARIABILITY

The challenge is to deter sectoral dominance while constructively addressing pressing concerns. The overall strategy breaks down sectors' dominance through diversification and flexibility. By developing options, policymakers need not succumb to delusions of certainty, but instead have choices. This is not a call for the good old days of simplicity or another, more populist, version of one best way. Instead, it is a call for skepticism, and it challenges the current intergovernmental system with suggestions that accept and build on the complex reality of uncertainty. Rather than channeling that complexity into specialized sectors, the proposals promote the system's variability.

A few examples illustrate the concept of variability and introduce its functions as a strategy. In general, the intergovernmental system tends to promote a single policy form—the prespecified program. Although legal and bureaucratic norms imply permanence, in practice, a policy may take different forms at different times. For example, low-income housing policy has changed from building monolithic projects to using marketlike vouchers. Similarly, environmental impact assessment report planning has changed from external, scientific advice to interactive decision making. These examples show that planning processes and policies can and do vary, and begin to suggest their potential for expanding choice.

The umbrella strategy of variability encompasses principles of decentralization, redundancy, and diversity. Each principle provides a different facet of variability. Decentralization implies breaking up or decoupling what was previously compacted and unitary into more flexible, responsive parts. Redundancy implies overlap and duplication to protect against the risks of a single, one best way (Landau, 1973) and to encourage alternatives. Diversity implies freedom of choice and of options.

Variability permits response to different problem conditions. This strategy fits with contingency theory (Alexander, 1996; Bryson & Delbecq, 1979; Lawrence & Lorsch, 1967), which means that the choice of an appropriate approach depends on circumstances. Although straightforward and obvious, this contingent approach contradicts bureaucratic norms and presumptions of certainty. Variability provides explicit, alternative ways to confront and deal with uncertainty. The idea that planning processes and public policies should vary in response to problem conditions invites challenges to premature programming and sectoral dominance. The strategy promotes both skepticism (Weick, 1979) and choice.

The strategy of variability also relies on the principles of democratic participation and diverse views (Webber, 1978). Such popular values seem obvious, but they nonetheless thwart bureaucratic norms and undermine narrow sectoral problem formulations and premature solutions. Broad participation challenges the "one best way" by encouraging multiple approaches, respecting differences, and using conflict and debate to elicit public choice.

The emphasis on increasing participation and expanding choice runs through a variety of proposals for privatization (Osborne & Gaebler, 1992), community empowerment and self-help (Boyte, 1990), and negotiation and consensus building (Innes, 1996; Susskind & Cruickshank, 1987). The emphasis on expanding options for self-interested choice—in opposition to the one right way—spans the political as well as scholarly spectrum.

Variability promotes constructive learning by encouraging adaptive response to differences. At the same time, planning processes and policies that are variable and thus easy to change are inherently cautious because they emphasize their tentative, experimental nature and because they make it easy to stop and adjust when something harmful starts to happen.

Contingent ways of conducting public policy can respond to uncertainty and decrease sectoral dominance. The contingent approaches described in the following sections all address variable expectations of governmental performance, set forth in Chapter 6, "Delusions of Certainty and Their Consequences for Expectations of Government," such as predictability, innovation, and accommodation of multiple preferences. The following proposals offer variable forms of policy, organization, and planning processes.

VARIABLE EXPECTATIONS OF GOVERNMENT

Expanding public expectations of government performance to consider conditions of uncertainty may be difficult. Resistance may stem from the norms of reliability and accountability (appropriate in conditions of certainty) embedded in our historical concept of responsible government. Attitudes cling to a simple model of government, whereas at least some actual governmental practice responds to current conditions.

In parts of some organizations, workers set their own goals and so sometimes tacitly, sometimes directly, confront uncertainty. For a direct example, medical researchers in a subunit of an epidemiology center try to discover the cause of a particular disease. For an indirect example, officials in the Environmental Protection Agency set goals and strategies for negotiating water quality issues in different physical and political settings. Some public policy thus begins to formulate expectations suitable to actual conditions of uncertainty.

The task is to help this embryonic tolerance of error, difference, and uncertainty grow into a fuller and wider spread recognition that different problem conditions call for different responses. To some extent, such a shift in public expectations of government can be encouraged by rhetoric. Instead of posing issues in terms of ignorance, ambiguity, and conflict, skilled politicians convert uncertainty and diversity into positive challenges or even into assets. For example, the 1996 welfare reform legislation is full of uncertainty and is cast in a broad block grant providing relatively unconstrained funding, which accordingly can respond to relatively unpredictable and diverse circumstances. Such grants may be characterized as giving cities and states their rights to make their choices.

In short, politicians and public servants can and do recognize uncertainty but are deterred by outmoded public expectations. The trick is to invite skepticism without the taint of despair. Good politicians can create constructive ways of promoting a range of expectations about government performance. They can exalt innovation and accommodation of diverse preferences as proper, up-to-date standards that are not merely palatable but actually upbeat.

VARIABLE FORMS OF POLICY

Just as decision makers have options in the ways they can frame expectations of government, so they have options in the ways they can frame policy. Often, however, opportunities for consciously tailoring the form of policy are obscured.

The legislative process tends to complicate the prospects for shaping the form of a policy. Usually a bill is drafted in its sector and promoted and lobbied through specialized committees by associated interest groups; then it makes its way over more general hurdles. Some bills even survive the complex lobbying, committee, log rolling, and budget maneuvering legislative process with their prescribed certainties intact. Other bills reflect compromise, ambiguity, and unresolved conflicts. Because they are the product of a particular set of actors in a particular historical, political, and economic context, the form such bills eventually take is hardly an accident. Nonetheless, their complex development makes them look haphazard.

To help clarify options obscured by such political processes, this section sets out bare-bones structural models of alternative forms of policy. Analyzing and illustrating the principles behind different forms of policy serves two purposes: first, to show that in practice, policymakers do make use of different forms of policy; and second, to help policymakers self-consciously design policy form to suit both circumstance and public purpose.

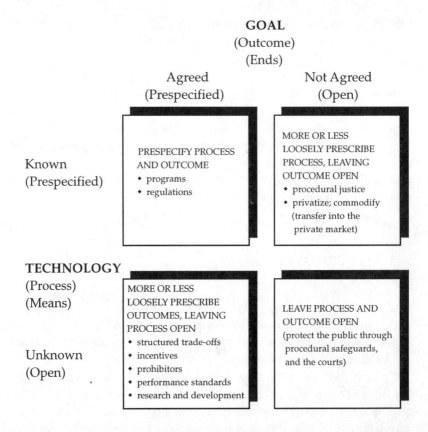

Figure 7.1. Variable Forms of Policy Associated With Variable Problem Conditions

A Framework of Variable Policy Form

To expose the principles underpinning policy form, Figure 7.1 shows how public policy processes and outcomes can vary.

The dimensions of the policy form framework present the two sides of the basic debate over whether policy ought to achieve particular goods or ensure a fair process. Some argue that the result matters; the outcome is crucial. Others argue that the result does not matter; the outcome is insignificant as long as it derives from a just process. Outcome advocates may say, for example, the purpose is to achieve clean air. Process advocates may say whatever mix of particulates ends up in the air is immaterial as long as firms follow a fair process of internalizing the social costs of polluting. Process corresponds loosely to technology and outcome to goal.

The uncertainty-certainty dichotomy has been recast into an assertive planning choice: to prespecify or leave open. *Prespecify* means simply that the policy is formulated in such a way that it sets particulars in advance, according to plan. *Open* means simply that particulars will be decided later, depending, perhaps, on different circumstances, opportunities, and preferences unknown to the planner. For example, development decisions in Alaska's North Slope Borough, a conservation district larger than California, are left open to preserve some flexibility in managing this vast unknown. Other conditions prompt cautious constraint. The risks of radioactive waste leaking into ground water are so serious, for example, that protection of underground repositories is scrupulously prespecified.

These divisions result in four basic alternative policy forms: (a) prespecified process and prespecified outcome; (b) open process and prespecified outcome; (c) prespecified process and open outcome; and (d) open process and open outcome.[1] Like most dichotomies, the distinctions blur at the edges. Process and outcome undoubtedly influence each other; what is prespecified and what is left open are usually a matter of degree. Nonetheless, the matrix offers a way of understanding not only the lines of debate but also the principles underlying many actual policies. Its simplicity may help policymakers keep these distinctions in mind, and so tailor legislation to problem conditions and public purposes as a matter of conscious design rather than merely as a by-product of politics and particular circumstances.

Prespecified Process and Prespecified Outcome

In prespecified process and prespecified outcome, policy prescribes a program. The policy produces a production function, identifying inputs and procedures and arranging them in the correct quantities in the correct place at the correct time. If the program is grounded in certainty over means and ends, it will suit the familiar bureaucratic model and its associated norms of predictability, equity, accountability, efficiency, and effectiveness. For example, storm drain capacities can be matched to local conditions by prescribed processes (specific guides) to achieve prescribed outcomes: no flood damage.

The operating principle is establishing rules. The rules can be written in legislation, guidelines, and manuals. They are connected to people and agencies through position descriptions and the complex intergovernmental arrangements described earlier.

Rules are also the basis for many regulations that straddle the boundary between public and private. Such regulations begin in bureaucracy but are carried out by countless individuals and firms. For example, the Clean Air Act amendments regulate the discharge of sulfur dioxide by allocating property rights and establishing a legal framework governing how they may be exchanged. Under this system, the government enforces rules to ensure that the objective of reducing acid rain is met, but private interests determine where and how the pollution cuts are achieved. Effective regulation, like effective programming, is embedded in known technologies and agreed goals. Thus, for example, the minimum federal standards on pollutants derive from studies on health hazards.

Open Process and Prespecified Outcome

Under open process and prespecified outcomes, the policy establishes the end but leaves the means of achieving it flexible. The extent of flexibility may be a matter of political choice or planning strategy.

Structured trade-offs is a policy form that is quite goal directed but still offers a safety valve. Such policy states that if one cannot or utterly refuses to comply with the rule, an option can compensate the public interest. In Berkeley, for example, code section 15-1.1 states that if a developer destroys a housing unit, the unit must be replaced on the original site or elsewhere in Berkeley. Given Berkeley's dearth of vacant land, prohibitive new construction costs, and political climate, the policy achieves its goal of preserving the status quo. Similarly, the federal Clean Air Act encourages states to prepare their own clean air legislation, but if a state fails to do so, the Environmental Protection Agency will prepare a plan for the recalcitrant state.

Incentives offer a form of policy that gives process a bit more latitude. Incentives provide "carrots" to encourage goal-achieving behavior. Households receive low-interest loans for insulating homes, for example, to reduce fuel consumption. Some cities and counties give developers a density bonus (which allows more dwelling units on the same parcel of land, and thus more profit) if they allocate a portion of their housing for the poor. Even after major reform, the U.S. tax code offers examples of incentives, such as mortgage interest deduction for home ownership. Moreover, much of the federal grant-in-aid system operates through incentives, offering state and local government funds to undertake a wide range of programs.

Some incentives may be cost-effective; others may not. But probably the key asset of incentives is providing choice. Politically attractive, incentives avoid the taint of coercion. Moreover, from a policy perspective, incentives offer planners valuable low-cost feedback on their implied production function. If many houses are insulated and fuel consumption decreases, if many developers create projects for households of diverse incomes, well and good. On the other hand, if towns or firms or individuals ignore incentives and continue to behave as usual, then the policymaker receives a clear signal: The implicit theory was wrong. Perhaps the other factors reinforcing the usual pattern are so strong that the incentive is trivial. In some cities, for example, density bonuses would be an inadequate carrot unless sweetened by national subsidies and strong neighborhood association support for low-income housing.

Prohibition is a policy form that provides a great deal of flexibility. Such policy sets a threshold or limit (a floor or a ceiling) on behavior, preventing public harm while otherwise being permissive. Thus, housing must meet various codes for electrical wiring and so on, but can do so in virtually whatever way an architect can imagine. During a severe drought, San Francisco Bay Area residents kept within their gallon limits per day, but each household used its water ration according to its priorities, with television news shows reporting how much ivy could survive on dirty dishwater. The federal government prohibits asbestos but permits a variety of other fire retardants.

Performance standards provide a policy form that is the positive counterpart to prohibition. Instead of preventing something bad (unsafe housing or wasting scarce water), performance standards define something good. Performance standards are the norm in the private sector. A contract specifies what is acceptable, and the producer makes it in any clever, profitable way that works. The purchaser has no interest in the method, as long as the product meets the contracted standards of performance. A city may contract for a garbage disposal service, for example, and specify only that so many tons of garbage be removed and disposed of in ways that comply with environmental protection laws, leaving the contracting firm or nonprofit agency operational scope. Performance standards are sometimes termed *output measures*. For example, in New York City, streets are assessed on a scorecard for cleanliness (Osborne & Gaebler 1992, p. 146).

A research and development (R&D) approach to policy may offer even greater scope for innovation. The general strategy behind most R&D policy is to give grants for disparate lines of study, all directed at a broadly defined goal. When U.S. dependence on oil seemed a crisis, grants were readily available for any sort of research connected to energy, especially to developing cheap alternatives to oil. The R&D form of policy actively encourages innovation at the expense of reliability. Sometimes the goal can be general innovation. For example, "Riverside, California set up a $100,000 seed fund, controlled by department heads to make small loans for new initiatives" (Osborne & Gaebler, 1992, p. 211). Whether single purpose or general purpose, an R&D goal is much more diffuse and expansive than the precise, narrow structures of performance standards. Furthermore, it downplays performance and instead rewards learning.

Structured trade-offs, incentives, prohibitions, performance standards, and R&D all offer flexibility over means. They vary in the degrees to which they focus on the end and leave the means open. The particular mix depends on political constraints and opportunities as well as on the policy problem.

For example, in its early days, the Community Action Program formulated its goal in terms of the war on poverty. Its decentralized strategy served President Johnson's political interests, bypassing the more conservative establishment in city halls to empower poor neighborhoods. This progressive structure coincided with the program's organizational design, which encouraged innovation by treating the problem, poverty, with action-oriented research and development. Later, under political attack and reduced funding, this expansive approach yielded to the city hall establishment, which treated the problem of poverty with the traditional provision of services. The Office of Economic Opportunity even had its name changed to Community Services. The problem of poverty had been converted to the problems of poverty as it lost its political urgency.

Prespecified Process and Open Outcome

This policy form establishes a process or procedure but leaves its result open. Its most familiar example is the U.S. legal system's due process. This policy establishes a fair procedure for judging guilt or innocence. In individual cases, plaintiff (or prosecutor) and defense disagree on desired outcomes but abide by prescribed procedures for the court's reaching a final decision.

In its more general form, this approach of agreeing in advance to abide by whatever emerges from an agreed procedure provides stability in many uncertain situations. In thousands upon thousands of legal cases, the outcome cannot be predicted, but due process ensures justice. Similarly,[2] elections follow scrupulous procedures; voters accept the winning representative, even if their preferred candidate lost. When individuals cannot predict whether or not they will benefit from a future decision (Buchanan & Tullock, 1962), they establish fair rules for reaching that decision, thereby prespecifying process while leaving the outcome open.

Collective public choice. Some public decisions must be reached as a whole because they cannot be easily subdivided. An individual must be judged guilty or innocent by all society, a president must be chosen by everyone, and the nation either uses nuclear power or does not. For such collective decisions, the U.S. system has developed complicated, multiple decision-making processes. The system defies simple majority rule and adds layers of decisions to induce caution. However complex these choices have become, the principle remains the same: Given uncertainty and the likelihood of multiple conflicting future preferences, preagreement on a decision process legitimates its result.

Numerous public programs use prespecified processes to make choices that are necessarily collective. The environmental impact assessment process of identifying issues and publicly debating them in advance of action provides a familiar example of adhering to a prespecified process for reaching collective decisions. The public agrees to tolerate a project's effects if the effects have been scrutinized, adjusted, and mitigated under the public's eye. Conversely, the public agrees to forego a project's benefits if its side effects are found to be overwhelmingly negative.

Other collective choices are made by boards and commissions (e.g., for the Community Housing Development Organization) for which the composition (e.g., one third must be lower-income residents or elected representatives of low-income organizations in the community) or procedure (e.g., mandatory public hearings) is prespecified. Undergirding all these complex processes is the basic governmental structure of elected representatives debating and reaching collective decisions in public meetings. This familiar republican form is an exemplar of a prescribed process with open outcomes.

Dividing public choice. Other choices need not necessarily be made as a whole. Instead, some may be broken down for constituents—firms, states, towns, households, persons—to make individual choices suited to their own interests. The guiding principle is to prescribe a fragmented process to allow multiple, diverse, even conflicting goals to be realized.

The most familiar example of putting this form of public policy to work is to use the market. This approach to prescribing process while permitting diverse outcomes can work when a public good is reformulated into a consumer good. In its ideal form, the market allocates resources efficiently and, in the process, provides self-regulation, innovation, adaptation to change, and freedom of choice. The market ideal neatly solves the policymaker's dilemma of trying to serve the public interest while recognizing that constituents actually have many different and sometimes conflicting interests.

In practice, however, the market may fail to attain the ideal. Some types of market failures are widely recognized, such as externalities (Moore, 1978). Moreover, because many problems enter the public domain precisely because the market cannot handle them, a strategy of using the market seems bizarre at first. Yet, where the boundary between public and private is placed is itself a matter of public policy.

Increasingly today, public responsibility shifts to the private sector. The postal department, for example, has become the quasi-private postal service. Now it attempts to operate according to market criteria by offering special features and trying to eliminate service to unprofitable areas. Once a legal monopoly, it now faces private sector competition, ranging from parcel delivery and cross-country overnight expresses to bicyclists in big-city financial districts. The strategy of shifting public activities to the private sector is sometimes termed *privatization* or *commodification* of what was previously understood as a unitary public goal.

A similar strategy is to give the poor the wherewithal to use the market themselves. For example, instead of housing poor people in large, stigmatizing projects in ghettos, housing vouchers permit poor people to find any decent housing unit in any neighborhood they like. The government pays the landlord the difference between what the household can afford and the market rent. Food stamps work on the same principle. Rather than constraining the market, the food stamps program allows producers to profit, innovate, and adjust to changing tastes while giving poor people more access to their markets.

A variant of this use-the-market approach generates competition among public service agencies (Levine, 1972). Instead of the conventional monolithic and monopolistic bureaucracy, this policy form intentionally establishes multiple agencies in an effort to create marketlike efficiencies and to give consumers choice. Following drastic funding cuts, an array of fee-for-service public after-school programs has created a veritable market to replace earlier free public programs. Many cities use community development block grant funds to fund nonprofit goods and service providers, which compete among themselves for the funding.

This marketlike strategy of prescribing process while leaving outcomes open to different preferences is feasible when a public good can be converted into a consumer commodity. In cases such as housing vouchers and food stamps, individual households can choose in their own interests. In general, privatization is effective when the task can be prespecified, performance is easily measured and evaluated, and contractors compete and can be replaced (Donahue, 1989, p. 97).

In theory, private enterprise can supply government with larger collective goods and services. This strategy puts government in the role of consumer, gaining the presumed benefits of competitive prices, flexibility, and choice. Rather than producing public services itself, government simply contracts with a private business for those services for a specified price over a specified time period. If the government is pleased with the service, it can renew the contract. O'Toole (1989) reports a neat set of six paired localities comparing public and private provision of wastewater treatment facilities. He found both public and private providers met the same high quality of water, but the private provider produced much faster because it had fewer change orders. Privatization resulted in reduced local autonomy, however, because the huge private investment led to a long contract. This conclusion suggests problems with scale.

In practice, this idea of private industry supplying public goods and services is often unworkable because of problems of scale. Services are either too bulky or too easily divided. Large services, such as utilities, create their own monopolistic district or are susceptible to collusive, even corrupt, ties to government. On the other end of the scale, small, easily divisible services, such as garbage collection, can be easily charged to individual customers. When public and private providers work in the same field, such as health care and electrical utilities, their costs and quality are roughly the same, with public utilities somewhat less expensive (Wilson, 1989, pp. 350-351).

The contracting-out strategy often works best where the government itself must be the owner, producer, or provider but can subcontract for components. For example, governments often contract out for social services and customarily solicit bids for private contractors to restore public buildings. Converting government into consumer takes advantage of the market's flexibility while providing collective goods and services. Some communities, "minimal cities," have adopted the Lakewood Plan, incorporating to achieve autonomy and control zoning, but contracting out for services to keep tax rates low (Miller, 1981).

Some collective goods and services allow diverse outcomes. Special assessment districts illustrate how the marketlike fee-for-service method can combine with the subdividing method. Instead of deciding on a citywide basis and paying for the improvement through the general fund or a bond, decisions and payments can be delegated to districts or neighborhoods. Thus, one neighborhood can charge itself for underground wires while another keeps costs down. In the San Francisco Bay Area, some cities, such as Berkeley, have paid extra to have the rapid transit below ground. This version of the open-ended policy form underscores the defects of its virtues. Subdivided fee-for-service offers some choice and diversity for collective goods. At the same time, it is patently inequitable on goods that are patently public. Thus, affluent neighborhoods have good public improvements and poor neighborhoods do not.

The community development block grant illustrates a special case of decentralized collective choices that compensate for unequal resources. Moreover, it shows wider advantages of the general policy of specifying process while leaving outcomes open. The authorizing act's primary objective is "providing decent housing and a suitable living environment and expanding economic opportunities, principally for persons of low and moderate income." But no certain, proven technology dictates how this objective might be achieved. Furthermore, the objective might mean different things in different geographical and political locations. Rather than prescribe what cities must do, legislation merely provides dependable funding and requires that a fixed percentage of it be spent in census tracts defined as poor. Although this spending requirement ensures accountability and focus on the general goal, its openness permits it to respond to diverse conditions and preferences.

Yet another way of using markets for public policy is creating market-like mechanisms to promote self-interested behavior, although still addressing a public goal. The tax on pollution (Schultz, 1977) illustrates how firms could have an incentive to reduce pollution while still deciding for themselves how and when to adapt their own technology. The Southern California Air Quality Management District is using a variation on that proposal ("Innovative Swap," 1994): All firms must reduce their pollution by a certain percentage, but if they exceed that amount, they may sell the surplus to other firms that may buy more polluting time to adjust their technology. All the firms have incentives to reduce their pollution as much and as quickly as possible.

Open Process and Open Outcome

General Revenue Sharing illustrates the most extreme form of policy openness: It verges on a policy void. Now defunct, General Revenue Sharing gave national money to states and cities to do with as they wished. Both process and outcome were open. Policies in this category are essentially laissez-faire. Many prohibitions constrain actual public behavior, however. Procedural safeguards such as public hearings and the courts offer some protection under even the most open policy.

Adopting Appropriate Policy Form

Policies need not be limited to a form prescribing both process and outcome. Such prescriptions imply certainty over both means and ends. Actual public policies can and do take many forms, implying some uncertainty and offering some flexibility.

The common sense behind these different policy forms can be hidden or distorted by political rhetoric and public perceptions of urgency. Scientists may vainly promote experiments they are convinced are low risk in the face of strong public resistance. When popular opinion is strong, policy cannot be relegated to the market. The marketlike alternatives of taxing milk companies according to the number of children they kill or fining residents for using too much water when none remains, for example, are politically unconscionable.

The point is threefold. First, the form of a public policy can vary and can be designed to address its particular problem conditions. Policy form can be a policy tool. Second, popular opinion and political rhetoric may constrain, or compel, the choice of policy form. Third, the policy form itself carries symbolic meanings. These in turn can be used for multiple purposes, some germane to the policy, others not.

The purpose in setting out the principles underpinning the variable forms of policy is to emphasize instrumental choice. Moreover, these principles expose not only the potential misuse of policy form but also its potential opportunities. To underscore these ideas, Figure 7.2 illustrates variable policy forms by applying the principles to a single issue: poverty.

Box A in Figure 7.2 prescribes both the means and ends of addressing poverty. It contains individual-directed benefits (payments) that are given equally, to everyone eligible, by entitlement, notably Social Security and unemployment insurance.

In Box B, policy forms encourage the general goal, reducing or ending poverty while leaving the process open. The new welfare reform block grant is intended to address poverty by moving people off welfare and into work. The performance standards (people off welfare) and prohibitions (e.g., end welfare after 2 years) are intended to encourage both states and welfare recipients to end welfare in whatever ways they can devise. The process is open to innovation.[3] At this writing, the most successful program, in Wisconsin, has had its welfare caseload drop by 60% (De Parles, August 24, 1997).

Public Policy in Uncertain Conditions

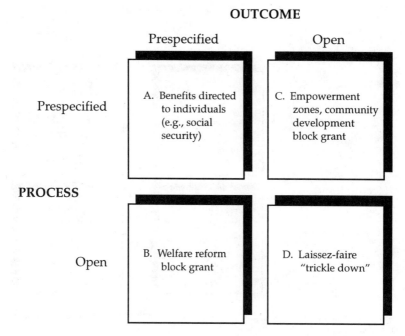

Figure 7.2. Illustrations of Variable Policy Forms Addressing Poverty

In Box C, processes are prespecified, but outcomes are left open. Perhaps the most familiar is the community development block grant. Its general goal includes "expanding economic opportunities principally for people of low and moderate income," and it requires that a certain percentage of funds be spent in low-income census tracts, but it does not specify what those activities should be. They can vary according to local physical, political, and economic conditions. A quite different example of a policy form that prescribes process but leaves outcomes open is voter registration drives or registration mechanisms targeted to poor people. Although the process of requiring registration for voting is rigorously prescribed, the outcomes of having more poor people voting are open to vast possibilities.

In Box D, neither process nor goal is prespecified. The resulting policy form is laissez-faire. When this policy form addresses poverty, it is sometimes called "trickle down," meaning as the economy grows, benefits eventually trickle down to the poor. The corresponding familiar saying is "a rising tide lifts all boats."

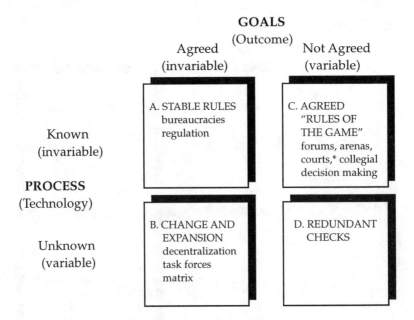

Figure 7.3. Variable Forms of Organization Associated With Variable Problem Conditions
SOURCE: Bryson and Crosby, 1992, pp. 81-117.

In short, a single goal of addressing poverty can be formulated in different policy forms with an array of effects. Although historical and immediate context shape policy into a particular form, that policy form may become a tool for furthering public goals.

VARIABLE FORMS OF ORGANIZATION

Although delusions of certainty tend to push policies into bureaucracies regardless of particular problem conditions, both theory and practice suggest alternative forms of organization.

Figure 7.3 sets out variable forms of organization that respond to different conditions of uncertainty.

Stable Rules for Known Technology and Agreed Goal

Conditions in Box A call for programming. When certainty prevails, rules make sense. Both outcome and process are predetermined because the rules are inherent in a desired, reliable function.

Rules are put to work in different organizational settings. First, a basic bureaucracy contains an entire production function, for a simple example, authorizing drivers' licenses. That takes place in a second, larger institutional system. In effect, granting licenses is only a part of the larger production function that relies on private drivers to prevent automobile accidents. Thus, public bureaucratic rules and regulation of private activities work in tandem. A third-level organizational setting, the intergovernmental system, employs multiple agencies—city police, county sheriffs, and state department of highway patrol[4]—to enforce compliance with the rules.

The example illustrates a simple structure and function grounded in a simple theory of traffic safety:

skilled drivers (certify) + good rules of driving (enforce) = safety

The rules derive from this production function. The state department of motor vehicles certifies that drivers are skilled through prescribed tasks (e.g., ensuring citizens' identity, testing their knowledge of driving rules, and testing their practical driving skills). Drivers are responsible for knowing and following driving rules. Enforcement officers from different levels of government are responsible for catching and deterring violators. This simple example illustrates the organizing principle of stable rules inherent in a known, desired function that is set in a single bureaucracy, regulates private activity, and operates in an intergovernmental complex.

Change and Expansion for Unknown Technology and Agreed Goal

The conditions in Box B, unknown technology and agreed goal, call for experimentation. The corresponding organization principle of change and expansion generates alternatives as means for discovering workable solutions. The matrix organizes activities along two important dimensions at the same time. It encourages change and expansion by fostering different combinations. A research firm, for example, organizes staff by discipline and project. Thus, an economist, transportation engineer, and lawyer tackle one project, while the same economist, a sociologist, a planner, and a social welfare specialist take on a quite different project in a quite different way.

A city might use a matrix organization to create different combinations of its housing, jobs, transportation, and open space policies to suit different neighborhoods. The U.S. Department of Housing and Urban Development (HUD) used a matrix organization to create a team of program specialists to help particular cities. A public housing representative, a code enforcement representative, and a model cities representative worked out some arrangements with Boise, while a different program team worked out quite different arrangements with Chicago.

Both illustrations show partial versions of a more complex matrix: federalism. Its dimensions are sector (vertical) and territory (horizontal). Figure 7.4 shows these three public matrix organizations.

A matrix organization's key assets are its flexibility and its capacity to diversify. These are the antithesis of bureaucracy's predictability and routine. By responding to diverse conditions and expanding problem formulations, the matrix organizational form promotes learning.

Agreed Rules of the Game for No Agreed Goal or Outcome

In Box C, a process is established to deal with conflicting goals and uncertainty over future preferences. Forms of policy that prescribe process but leave outcomes open imply their corresponding organization principle: agreed "rules of the game." The organization may take the form of markets, courts, legislatures, or any kind of bargaining arrangement. Participants may be private interests, government representatives, or both. The organization simply provides a structure within which particular agreements can be reached according to these prearranged rules.

Redundant Complexity for Unknown Technology and No Agreed Goal

The conditions in Box D, unknown technology and no agreed goal, are open to the point of chaos. The compatible policy of laissez-faire seems to defy public organization but calls for some sort of protection against anarchy. Conditions of uncertainty over both means and ends create a tension between the appeal of unrestricted opportunity and the fear of chaotic conflict and disintegration.

Public Policy in Uncertain Conditions   135

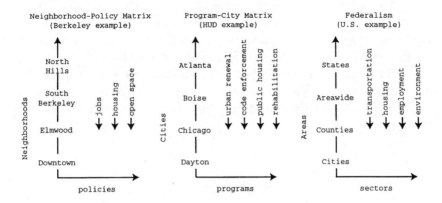

Figure 7.4. Public Matrix Organizations

The organization principle for coping with this tension and uncertainty is redundant checks (Landau, 1969, 1973). Because the intergovernmental system is so complex, it can afford to leave some issues to laissez-faire. On the one hand, many choices can be left without policy because their sheer numbers counteract each other and create a sort of shock absorber. On the other hand, so much of the system operates fairly well in less chaotic conditions that it tends to muffle problems of uncertainty over means and ends. Redundant complexity creates a protective context that permits some parts of the system to appear laissez-faire.

This conclusion neither condones the effects of laissez-faire policy nor ensures system equilibrium. People and places may be hurt, redundancy may reinforce privilege, and forces may interact in ways that change the system, perhaps abruptly. The conclusion simply emphasizes that even laissez-faire does not rest in a policy vacuum.

Figure 7.3 presents categories that are deceptively tidy and static. Actual organizational forms are hybrid and dynamic. Members of a task force, for example, may come from bureaucracies, collegial groups, political cabals, or competitive enterprises. A single organization may mix bureaucratic and nonbureaucratic organizational forms, and its subunit's organizational form may change over time. For example, one part of a regional Environmental Protection Agency (EPA) office may be participating in intense interagency and intergovernmental negotiations over water quality standards for an estuary, whereas another division down the hall routinely monitors point source pollution, and yet another division facilitates community planning for a toxic cleanup. Organizational form varies over time as workable solutions gradually become routine, and routinized, while other routines unravel (Weick, 1979). Variable organizational forms describe both categorical types and an ongoing process.

VARIABLE FORMS OF PLANNING

Traditionally, planning concerned both means and ends and often assumed both were known. This professional legacy reinforces current institutional predispositions to certainty and rational planning (Dalton, 1986). Thus, planners are prone to misconstrue problem conditions (Bolan, 1969; Cartwright, 1973; Rondinelli, 1973) and to use inappropriate traditional processes that lead to ineffective or harmful results. By matching planning processes to actual problem conditions, however, planning offers the promise of reducing uncertainty. In this way, planning processes can be understood not as static givens but instead as both contingent[5] (Alexander, 1996; Bryson & Delbecq, 1979) and instrumental.

Planning for Known Technology, Agreed Goal

These conditions are the basis for planning theory's traditional doctrine, which has been under attack (Alexander, 1986; Innes, 1995). Critics argue that no planning problem has a known technology and an agreed goal, that is, a clearly identified problem with a proven solution. Although it is easy to disparage conditions of known technology and agreed goal as trivial nonproblems, these conditions nonetheless characterize many public activities. Examples such as water and sewer sanitation to reduce epidemics suggest that the situations were at one time public problems that have subsequently been "solved" but require recurring attention.

The planner knows the goal and sets about matching an effective technology to accomplish the goal. This rationalist approach—which requires a known technology (means) and an agreed goal (end)—is central to the traditional planning doctrine. A land use planner aiming at preserving a town's single-family residential character, for example, would establish low-density, R-1 zoning. Other cities might chose to zone for more concentrated, pedestrian-friendly, transit-oriented development. Both use the rational model.

Numerous subdoctrines elaborate the rationalist planning model. Perhaps the most pervasive is that of expertise. Expertise may be highly specialized and almost scientific, for example, expertise in analyzing housing markets; or it may be a more generic skill in selecting effective means to reach known ends. In this view, planners assume the roles of programmer, standardizer, rule setter, and regulator. Thus, planners might establish procedures for reviewing grant applications or assigning clients to caseworkers; other planners may write housing codes regulating private construction.

A more explicitly technical cast to the planner's presumed expertise in matching effective means to a given end concerns setting and ordering tasks in the "best" way. Associated planner roles then are scheduler and optimizer. Similarly, the planner as analyst is thought to be expert in choosing the best alternative, customarily through cost-benefit or cost-effectiveness computations. So, for example, a planner could tally the comparative land, construction, interest, maintenance, and pollution costs of a dozen alternative garbage disposal proposals to identify a city's best choice.

Another set of planner roles addressing known technology and agreed goals is administering. The certain production function of known tasks and procedures to produce a desired public service casts the planner in the roles of manager and evaluator. The general administrative purpose of ensuring smooth, efficient, effective, equitable, predictable operation suits conditions of known technology and agreed-goal.

Planning for Unknown Technology, Agreed Goal

Public commitment to solve a pressing problem (agreed goal) with no proven solution (unknown technology) requires a search for a workable solution. This search for technology is a reasonable, appropriate response to the problem conditions. Two related sets of theories and corresponding planning roles offer different approaches to discovering useful means.

The first set of theories presupposes that people's capacity for knowledge is inadequate and proposes ways to act given this inherent incapacity. Such theories include pragmatism (Blanco, 1994; Kaplan, 1961), incrementalism (Hirschmann & Lindblom, 1962, p. 211), "muddling through" (Lindblom, 1959, pp. 79-88), satisficing rather than optimizing behavior (Simon, 1976), and problem-focused (Cyert & March, 1963) rather than comprehensive, goal-oriented search. In this general approach to uncertain means, the planner behaves cybernetically (Steinbrunner, 1974), trying something, receiving feedback from the environment, and then making further modifications in response until the situation becomes tolerable. In this sense, planning is organizational learning (Benveniste, 1989) and organizing planning to learn (Michael, 1971). Planning in these conditions is as much about learning as action, and is best conducted through multiple approaches that can be easily reversed (La Porte, 1975, pp. 348-349).

This approach seems the opposite of the classic rational response to conditions of certainty. Rather than try to predict in advance as many consequences of a potential action as possible, the planner takes a plausible action first and then waits to see what consequences actually occur. If the consequences are acceptable, then the action is deemed workable and may be repeated. If the consequences are unacceptable, a new variation is tried. Pragmatic planning aims at finding something that works. Overall, this process tends to generate instrumental knowledge and thus reduces uncertainty. In this adaptive mode of planning, the planner takes on the roles of pragmatist and adjuster.

Some cities' zoning laws illustrate incremental pragmatic planning of land use and urban form. The zoning ordinance is often slightly amended to adjust to market forces, allowing for particular types of commercial uses: curbing an excessive number of bars, for example, or even establishing special use districts to retain neighborhood character in the form of a cherished soda fountain. At the same time, individuals are granted variances—even more finely hewn, plot-by-plot exceptions to zoning to bend to particular circumstances.

This zoning illustration shows pragmatic, adjusting planning. It reverses the example of known technology and agreed goal, when the planner forthrightly zoned single family to maintain the town's low-density residential land use. In this second example, the goal of reasonable land use is more general, and the method is less clear-cut. Frequent feedback and adjustment suit a view of planning as accommodating the future rather than designing it.

A similar pragmatic process emerges when more is presumed to be known. For example, urban renewal was thought to be a workable way to revive decaying inner cities. Accordingly, it was prematurely programmed and all but inevitably led to unexpected, disappointing results, notably enormous delays, expense, and destruction of neighborhoods (Gans, 1962) and hundred of thousands of housing units (National Commission on Urban Problems, 1969).

The program was reprogrammed repeatedly in bits and pieces. Some changes reduced the program's time, expense, and cost; some required citizen participation; some encouraged rehabilitation rather than clearance; and some curtailed dislocating people. Through this iterative process, the program gradually learned to be less destructive. (Each change was treated as if it were certain, however.)

A second set of planning roles derives from a more scientific tradition that treats uncertainty (the unknown technology) with conscious experimentation to discover a workable solution. In contrast to the first group, this set presupposes that command of the necessary knowledge is possible and proposes ways to attain it. Planning theorists in this school stress planning as learning and as systematically applying strategic variations (Dror, 1968; Michael, 1971; Rivlin, 1971). In this way, ignorance is confronted directly. In comparison with the pragmatic approach, the search for knowledge is more explicit and proactive; it focuses on methods to generate knowledge.

Although few pure experiments can be conducted in public policy, planners aim for scientific norms in a variety of ways: research and development, pilot programs undertaken in selected varying conditions, alternative methodologies, phasing alternative developments to defer choice until knowledge is clear, and establishing conditions for learning as part of the planning process. These attempts may generate either formal causal knowledge or simple "how-to" knowledge to reduce uncertainty. This approach to planning stresses the roles of learner, experimenter, and innovator.

A child abuse study offers an example of this purposeful learning process. Planners established pilot programs with several systematically varied ways of helping parents and children, and included extensive program monitoring and staff and client interviewing. Through several years of operation, they could distinguish useful and useless methods of preventing child abuse. In addition to this practical new information, planners discovered an important theoretical relationship between abuse and retarded children. By thinking and acting systematically, to coin Rivlin's (1971) title, planners confronted the child abuse puzzle directly.

Planning for Known Technology, No Agreed Goal

Planners frequently face confusion and disagreement over ends. Resolution comes through bargaining-like processes that depend on the exercise of power in a context of human values. In these expressly political conditions, the planner undertakes a range of planning roles and processes to accommodate conflicting goals.

The prototype form of decision making in these conditions is bargaining. Consensus building (Innes, 1996; Innes et al., 1994; Susskind, 1981; Susskind & Cruickshank, 1977), communicative practice (Forester, 1993), transactive planning (Friedmann, 1987), and facilitating (Archibald, 1970) encourage communication to emphasize common ground and shared goals. On the other hand, mediating (Susskind & Ozawa, 1983) and conventional bargaining assume differences, and reach agreements through trade and concessions.

Advocacy planning (Checkoway, 1993; Davidoff, 1965) emphasizes conflict still more. Like lawyers, planners take their clients' point of view, argue, and set forth proposals in their clients' interests. Advocacy planners give disadvantaged groups a voice in the hope that a wide range of competing arguments will prompt response to more interests. Promoting more active (Friedmann, 1973) and diverse (Webber, 1978) participation also incorporates humanitarian concerns in attempts to remedy bargaining disadvantages caused by unequal power. The constitution-writing method for accommodating multiple preferences aims at structuring the rules of and chips for bargaining in such a way that all affected interests can participate fairly.

For example, planners structured a liquefied natural gas terminal siting problem[6] in terms of an auction. Ordinarily, sectors place such risky

Public Policy in Uncertain Conditions 141

public works in poor, politically weak communities that have little power to resist the imposition. In this case, planners changed the usual rules of the game to give communities some leverage. These communities bid for the undesirable terminal by saying what they would accept as compensation for the risk. The "winning" community received a package of public goods in exchange for the terminal. Given the structure to bargain, the disadvantaged were able to gain some benefits even though they were still disadvantaged compared with affluent communities.

Another quite different approach to known technology and no agreed goal is to divide issues into subsets to avoid conflict and the need for resolution. A planning role compatible with this approach is resource allocator. The planner skirts uncertainty over goals by dividing resources (however weighted) among the different, often competing, goals. For example, some planning department staff may work on development while others work on preservation.

Planning for Unknown Technology, Unknown Goal

Uncertainty over both means and ends is extraordinarily disturbing and unstable. Planners frequently confront these conditions in practice (Rittel & Webber, 1973, pp. 155-169; Schon, 1983). The task of creating some form of order can be termed *problem finding*.[7] Uncertainty over both means and ends demands that the planner articulate the issue. The way the problem is formulated must be so compelling and intelligible that it can provide an acceptable reference point for subsequent attempts at resolution.

Such problem finding—that is, problem identification or articulation—may require problem reformulation, casting the problem in such a new light that people can agree that it is the right problem to tackle. Reformulation in this sense entails insight into both the nature of recalcitrant problems and political forces to ensure agreement over the problem. When reformulation is successful, uncertainty over the goal has been reduced, and conditions are simplified so that attention can focus on technical aspects of problem solution. In effect, the problem is thrust left, into Box B in Figure 7.5. (unknown technology, agreed goals). For a time, Lyndon Johnson resoundingly articulated and moved the problem of poverty out of Box D's chaotic depression into Box B's directed energy.

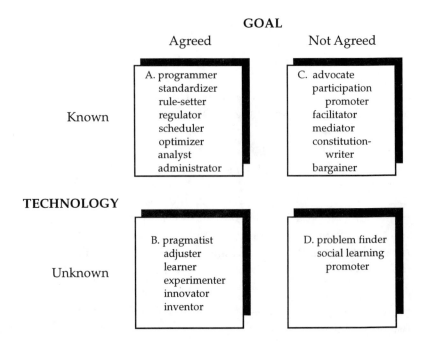

Figure 7.5. Planning Roles Categorized by Problem Conditions

Problem finding may require sifting through and articulating conflicting, vague goals to show how debate could focus on alternative goals with effective technologies. In this way, the problem is articulated as conflict. For example, the problem of crime may be found to be a symptom of widening gaps between the haves and the have-nots. Again, clarifying the issue requires insight into both the substance of the problems (to pose solvable alternatives) and political preferences. This kind of problem finding results in agreement not on the problem but on the choices to be debated and reconciled. When this sort of issue clarification is successful, it reduces uncertainty over technology, and conditions are simplified so that attention can focus on accommodating multiple goals. In effect, the problem is thrust upward into Box C in Figure 7.5 (known technology, no agreed goals).

The planning process to encourage problem clarification is promoting social learning. As discussed in Chapter 6, "Delusions of Certainty and Their Consequences for Expectations of Government," social learning can emerge from a variety of interactive planning processes. The planner's job is to elicit, structure, and facilitate participation in ways that are fair, open, and productive (Bryson & Crosby, 1992; Forester, 1989; Innes, 1995).

Summary of Planning With Respect to Variable Problem Conditions

Figure 7.5 categorizes planning processes associated with different theories of planning by the problem conditions they address.

Figure 7.5 implies that planning processes can be understood as addressing different conditions of uncertainty. Thus, planners must assess the actual conditions of uncertainty that characterize the particular problem they are confronting and then select a style of planning that suits those conditions. By tailoring planning to real-world conditions, the planner is acting contingently.

Intergovernmental System Influences on Planning Practice

The intergovernmental system tends to constrain everyday planning practice. Sectoral dominance, specialization, and artificial creation of certainty nurture a deceptively instrumental rationalism. Thus, even where technology and goal may be seen as uncertain, planners and managers try to program.

Sectoral specialization also reinforces the power of presumed experts. When surprises, political demands, or external crises threaten a sector's presumed known technology and agreed goal, remedy is sought within the specialty. Adjustments are marginal and must be acceptable to both the sector's professionals and its supporting legislative committees, clients, and vested interests.

Thus, the intergovernmental system tends to overemphasize rational planning and suppress experimental, innovative, and inventive styles of planning. Because the task of posing program options is usually consigned to specialized experts, the potential for a broad range of alternative ideas is usually lost. Moreover, specialization tends to blind experts to the substantial kinds of reformulation demanded by uncertainty over both technology and goals. Thus, although the intergovernmental system allows minor technical adjustments, it deters innovation and major problem reformulation.

Because the intergovernmental system avoids posing trade-offs across sectors, logically it would seem to preclude planning processes that address uncertainty over goals. Nonetheless, in practice, the system demands bargaining in three ways.

First, interactions within sectors over technological shifts entail negotiations and consensus building. Variations in technology necessarily alter power and resources among the levels within each sector. The levels jockey alternatives back and forth, emphasizing their differences and competition. Arguments may enjoin ideas about federalism to bolster a particular position. For example, the national level recently shifted an expensive service, welfare, onto the states, partly under the guise of states' rights.

At the same time, a sector's levels are bound together in a specialized reference group; they are accustomed to collegial consensus. In such collaborative modes, officials may work out acceptable mutual adjustments (e.g., over new functional literacy standards) at professional association meetings. Depending on the issue, timing, and setting, sectoral interactions may appear more conflictual or more consensual.

Second, adapting relatively mature sectoral technologies to area conditions and preferences entails some bargaining that may be either explicit or tacit. For example, a local transit planner's journey-to-work data may need to be adjusted to accommodate a regional transit planner's intention to link the locality to the metropolitan freeway network. The planners may dispute each other or merely adjust some figures without a bargain being obvious to either party.

Third, and most significant, area resistance to sectoral impositions demands effective styles of bargaining. Although bargaining is skewed toward sectoral interests, programs change as a consequence of these bargains. Moreover, area articulation of the costs of sectoral impositions is the major check against the unbridled sway of special interests. The intergovernmental system encourages bargaining styles of planning such as advocacy, negotiation, consensus building, and, if necessary, resistance.

Because of sectoral dominance, specialization, delusions of certainty, and premature programming, what customarily goes by the name of planning often fails. Results rarely conform with intentions. Programs incur strange, sometimes harmful, unforeseen consequences. Because the intergovernmental system predisposes planning to rationalist planning often based on false certainty over means and ends, and because it deters innovative and problem-finding styles of planning, the system hampers planners' capacity to respond to actual problem conditions. As shaped and constrained by the intergovernmental system, planning practice wears a rational veneer. But because this posture is often inappropriate, it is unreasonable.

Challenging System Constraints on Planning Practice

The preceding sections advocate tailoring planning processes and policy forms to actual, particular problem conditions and then explain why the intergovernmental system hampers doing so. This contradiction points to the powerful momentum of the intergovernmental system as it has evolved. Sectors developed through specialization in response to political demands; buttressed by legislative committee, professional, and economic interests, they have entrenched their power. Moreover, political processes seem incapable of curbing sectoral dominance, and area interests are weak checks.

Bucking such a powerful system seems so difficult that planners and policymakers may well despair. Small wonder that most planners retreat into the technical specialization that is consonant with system forces. But when they do so, they generate still more incremental, untested reforms. These stimulate governmental interactions, and so perpetuate the system.

Thoughtful planners and policymakers might nonetheless try to overcome some of the system's constraints on practice. In addition to questioning their own expert-based certainty regarding both goals and technology, planners and policymakers can reduce the detrimental effects of institutional delusions of certainty in several ways.

First, every time surprises or political demands reveal a sector's technological immaturity or conflict over goals, response should not be left to the presumed professional experts whose program produced that surprise. Instead, the stimulus should be treated for what it is: a symptom of uncertainty. Inviting a wide range of ideas[8]—from experts in other fields, people affected by the program, academics, as well as professional experts—should achieve the full, open search for a solution that conditions demand. At the same time, widening discussion beyond the sector's professional peer group should free deliberation from the constraints of the group's tacit assumptions and shared perceptions.

By the same token, expanding participation may require extra efforts to ensure that diverse groups can communicate effectively with each other. Forester (1989) developed standards for communicative practice that try to ensure that communication is undistorted and fair and focuses attention constructively.

Expanding participation beyond the usual group is liable to expand conflict and thus postpone agreement on a potential solution until one can be fashioned that is acceptable to the wider group.

Although slowing the initial process, a variety of views encourages problem reformulation, greater innovation, and eventual responsiveness. Innes et al. (1994) found that these processes can begin with high levels of mistrust and skepticism and, nonetheless, work through to consensus. The process entails developing joint knowledge and building social and political capital (Gruber, 1994). These kinds of diverse, sometimes conflictual, participation take place in a "shared power world" of forums and arenas (Bryson & Crosby, 1992) and ad hoc arrangements.

Second, when a potential solution is agreed on, it should be treated as a working hypothesis (Landau & Stout, 1979) rather than as a certain technology. Recognizing the proposed solution's tentativeness indicates that it ought not be institutionalized into bureaucracies, rules, and procedures (de Neufville & Christensen, 1981). Instead, the proposed solution should be tested and closely monitored for its consequences. Viewing the proposed solution as a working hypothesis allows openness in examining all consequences, because even unfortunate results are seen as useful information rather than as infractions of rules (Landau & Stout, 1979). Similarly, such a viewpoint encourages frequent modifications, because the aim is neither pure science nor bureaucratic predictability, but rather finding something that works.

Moreover, recognizing the experimental nature of the proposed solution manifests its riskiness.[9] Confronting risks openly should induce caution. One important way to deal with risk is to involve the people to be affected by the experiment. They may be able to identify some risks that may elude the experts. They may thus anticipate some adverse consequences that can be corrected before the hypothesis is tested, thereby reducing harm and increasing chances for a workable solution. By recognizing that the proposed solution is a working hypothesis rather than a certain technology, experts should cede some of their expertise to the people who are affected.

Jointly, the presumed experts and those immediately affected can deem a hypothesis to be a workable solution. It will be deemed workable if it furthers sectoral policy while helping a particular place and particular people. If the original idea fails to satisfy them, they can keep adjusting until something does. This perspective is immediate and specific.

Cumulative effects of this tentative, small-scale, short-term strategy may seem out of control. Nevertheless, two processes provide some overall guidance. First, promising workable solutions connect and transcend particulars through many lines of communication (including the associated sector). Eventually, these small workable solutions may build a reliable technology. Second, the caution and care underlying this approach encourage every particular activity to be tentative. Thus, collectively, these activities are unlikely to foreclose future choices. Recognizing uncertainty on a small, particular scale helps to keep the larger system open.

Urban renewal illustrates premature consensus and premature programming. Despite the pain and losses its subsequent program trials imposed, today's evolved form of urban renewal illustrates constructive processes for dealing with uncertainty openly. Urban renewal has become increasingly small, cautious, and conserving, involving more rehabilitation and less clearance, more housing and less commerce, more citizen participation and less central, end-state planning. This shift responds not only to escalating costs and human demands but also to conditions of uncertainty.

Communities now use block grants to tailor activities to their unique physical, social, and political conditions. Thus, in a sense, there is no longer a national program of urban renewal[10] but instead an array of local programs. Communities that choose neighborhood revitalization (note that this current term has the old theme, but in a smaller, more cautious voice) involve residents who shape activities to their own concerns and monitor those activities closely (Ahlbrandt & Brophy, 1975). Over time, many cities have developed variations on neighborhood strategies and exchanged ideas: A workable technology is emerging. The decentralizing course from urban renewal to neighborhood revitalization shows an openness to uncertainty that encourages not only caution but also learning.

SUMMARY

Specialization and delusions of certainty predispose the U.S. governmental system toward rationalist planning, programs, and bureaucracy. Yet many questions of public choice are clouded in uncertainty. The message on conducting public policy has been threefold. First, the chapter merely described a complex reality. Much of government actually operates with

- expectations of uncertainty (such as accommodating conflicts),
- forms of policy for uncertainty (such as open-ended legislation like community development block grants),
- organizations for uncertainty (such as ad hoc problem-solving task forces), and
- planning processes for uncertainty (such as bargaining).

Despite strong popular and systemic predispositions to certainty, parts of government act in ways that directly confront and cope with uncertainty.

Second, expectations, policy form, organizational form, and planning process can vary. They need not be simply accepted as given. Decision makers have an array of choices available. Because policy form, organizational form, and planning process can be purposefully chosen, they can be matched to different problem conditions. Thus, the way public policy is conducted can be used to further policy.

Third, this variability should be consciously exploited to forestall premature programming and premature consensus. By challenging sectoral dominance and conducting public policy to respond to actual uncertainty, officials can guard against some harms, save some time and money, and keep more options open. If uncertainty is the source of many public problems, it can also be the path to their resolution.

NOTES

1. Stoker (1991, p. 95) has developed a related three-by-three matrix with distribution of public authority on one dimension and action sector (government, joint, market) on the other dimension. The first box is the same as the first box in Figure 7.1. The open dimension here corresponds somewhat to Stoker's "diffuse" and "market" categories.

2. The processes, procedures, and flexibility regarding the outcomes are quite different in legal cases and voting. Nonetheless, the general principle of trusting the procedures to achieve a just outcome is the same.

3. The welfare reform legislation block grant and performance standards are policy forms appropriate to the problem conditions of agreed goal, no known technology. Other aspects of the legislation (e.g., premature rules), however, are liable to cause harm and to impede learning, however.

4. The national level participates not through police but through the Department of Transportation's Office of Highway Safety.

5. *Contingency management* refers to organization theories that advocate applying different management practices depending on (contingent on) conditions.

6. The liquefied natural gas terminal siting problem and solution were developed in a studio course, led by Lawrence Susskind and David Dowall, at the University of California, Berkeley, in the spring of 1978.

7. This is one of Melvin Webber's 20-odd "notions of planning" presented in introductory lectures, city and regional planning, at the University of California, Berkeley. Donald Schon (1983) elaborates the idea (see especially pp. 40-41).

8. One method for generating a broad range of ideas is to ask one person to identify someone with extremely different views than himself or herself, and ask the next person to identify someone with extremely different views than himself or herself, and so on (Guba & Lincoln, 1989).

9. Premature programming of uncertain technologies also entails risks, but they are disguised by the mistaken presumption of certainty and imposed on the disadvantaged (Marris, 1996).

10. The original idea resurfaced during the Carter administration, however, under the name of "urban demonstration action grants" and, more recently, in enterprise and empowerment zones. In addition, of course, old urban renewal projects continue their long road to completion.

CHAPTER 8

CONCLUSIONS FOR THE INTERGOVERNMENTAL SYSTEM

Federalism is a form of government capable of dealing with uncertainty. Development has fused the U.S. intergovernmental system's checks into sectoral dominance and delusions of certainty, however. This chapter begins by summarizing the book's themes of sectoral versus area government and certainty versus uncertainty. Next, the chapter distinguishes the essence of federalism from its operations. It concludes that weak territorial checks need not doom us to futile attempts to reinvent the past or to abandon the values of federalism. Instead, by shedding delusions of certainty, public policy can adapt and reinvigorate federalism's promise to cope with the uncertain challenges of the future.

SUMMARY OF THE ARGUMENT

Sectoral Versus Area Governments

The popular and traditional conception of government is rooted in territory. Thus, area government—city, county, state, national—is conventionally understood as government's base. Perhaps when the constitution was framed area governments could represent discrete, autonomous interests. But in today's advanced political economy, area governments are interdependent, and their mobile constituents participate in activities that spill over territorial boundaries.

Sectors linking federal, state, areawide, and local government agencies by specialty overcome traditional area divisions. Agencies within each sector share the same mega goal, professional perspective, and interdependent set of tasks and procedures. Each chain of agencies is buttressed by legislative committees and private, vested interest groups serving their shared specialty. Sectors act independently of one another, each developing and modifying its programs according to the interplay of politics and its task technology. A sector applies its programs to real places and people, represented by area governments.

Collectively, many different sectors impose their products on an area in a way that is necessarily disjointed. Because area government cannot predict, much less coordinate, these sectoral effects, it customarily relegates the area's response to the specialized agency within the respective sectors. Thus, public policies are formulated within specialized sectors. Area government is weak in the face of sectoral strength. Because of its diverse and fluctuating constituencies, area government can rarely form a coherent policy (and even then, not one that can last). Instead, an area is left with the choice of accepting, adapting to, or resisting a sector's impositions.

This pattern deters debate over mega goals, skews democratic access away from the general public toward special interests, and shunts deliberation away from larger questions toward narrower issues.

Certainty Versus Uncertainty

Uncertainty is endemic to governance, and yet it is rarely acknowledged. People like to believe that the government knows how to do things and that what it does is good for everyone. Or, more frequently today, people believe with certainty that government is inept. "Ronald Reagan's legacy was to turn American suspicion of politics into contempt for the whole government" (Wills, 1996, p. 30). Either way, legal, political, and administrative norms are predisposed to certainty. Laws clarify rights and wrongs; politicians offer platforms with confidence; administrators apply rules for reliability. Officials articulate tasks with assurance that means are known and effective and goals are widely accepted.

But the tasks of government are far from certain. People may be baffled by technologies, in conflict over goals, or uncertain about both means and ends. Moreover, current uncertainties expand into the future. Pressing problems, political preferences, and technological threats and opportunities are not susceptible to long-term prediction. Thus, much government action addresses real conditions of uncertainty.

The contemporary U.S. intergovernmental system does not reconcile the contradiction between actual conditions of uncertainty and predispositions toward certainty. Patterns of specialization frame issues and segment policies into a structure that perpetuates delusions of certainty and leads to premature programming. Sectors treat issues as if uncertain technologies were proven effective and goals were widely accepted.

The costs of delusions of certainty and sectoral dominance bear repeating:

- proliferating programs and agencies,
- damages of sectors' errors on places and people,
- formulation of public policy by special interests,
- skewed democratic access, and
- deterrence of debate over mega goals.

Delusions of certainty suppress the importance of adjusting, learning, and challenging. If real conditions of uncertainty were acknowledged, these crucial tasks could be addressed openly and directly, as set forth in Chapter 7, "Conducting Public Policy in Conditions of Uncertainty." Moreover, government would inflict less harm on people, places, and the governmental system.

CONTRADICTIONS OF FEDERALISM TODAY

Federalism was never simple. What emerged from complex historical circumstances and public deliberations over profound dilemmas was government in which citizens simultaneously belong to two jurisdictions. The framers invented U.S. federalism to protect both union and diversity and to guard against both tyranny and anarchy. Federalism arose in an earlier era to deal with enduring problems. Its complex form suits complex issues and endows it with a capability to vary, and thus the promise of coping with uncertainties.

But the public and politicians yearn for simplicity and certainty. Many cast this desire into a mythical past government dressed in the illusion of simplicity. To some, federalism seems simple because it was invented in an era without today's territory transcending cars, multinational corporations, and the Internet. In the framers' era, territorial divisions contained important social, political, and economic distinctions. These clear divisions may make the traditional concept of federalism appear simple.

Confusing federalism with territorial divisions and simplicity leads some to believe that if they could restore government's territorial divisions, they could achieve simplicity. Indeed, much of what is referred to as new federalism and the "devolution revolution" are proposals to transfer national and intergovernmental decision-making authority to states. Through the development pressures of its political economy, however, the U.S. government has outgrown its original form while preserving its federal name. Development, specialization, special interest policy formation, and sectoral dominance have eroded the U.S. government's territorial basis and vitiated cosovereignty.[1]

The rest of this chapter wrestles with the apparent contradiction of federalism by disentangling the concept from prevalent practice to salvage the promise of federalism for the future. To begin, it assesses current operations in terms of the three dominant models of federalism. Next, it explores federal principles of multiple governments, checks, and conservatism. It concludes with ways federalism can work for the future.

Current Intergovernmental System Considered in Terms of Three Prevailing Models of Federalism

Just as the current intergovernmental system has outgrown its initial design, so it overwhelms theoretical models. None of the three prevailing models of federalism captures the complexity of today's government. Yet each conveys an important aspect of system dynamics. The three models, discussed in Chapter 2, "Competing Theories of the U.S. Intergovernmental System," are dual (autonomous responsibilities), multicentered (shared responsibilities), and functional (hierarchical responsibilities).

Clearly, the sectors' power substantiates functional federalism. But the sectors' multiplicity and proliferation belie this single, coherent model in two respects. First, the system segments sectors and discourages cross-sector choices. Second, sectors' proliferation not only elaborates technologies but also increases plural centers of power. Moreover, the agencies within sectors are interdependent, not hierarchical. Thus, functional federalism captures the power of vertical integration but misses the intergovernmental system's complexity and dynamics.

The intergovernmental system's capability to continue to generate new programs, decision points, and actors displays some key features of multicentered federalism. Moreover, because sectors' problems and technologies vary and because some are in flux, they afford a variety of leverage points, multiple cracks, for change.

Nonetheless, intergovernmental dynamics skew democratic access toward sectors and accordingly toward special interests. In this light, traditional dual federalism's rallying cries of "states' rights" and "home rule" are not mere anachronisms but the power to resist sectors' impositions. In one sense, home rule simply rejects a sector's choice for one particular area of government. In a larger sense, however, widespread area resistance informs a sector that its current technology, or perhaps even current formulation of its goal, is unacceptable to a broader-based polity.

Thus, each of the three contradictory theoretical models is partly and importantly telling but incomplete. To the extent they all contribute to understanding a more complex system, they can be reconciled.

Enduring Principles of Federalism and Current Operations

The principles of federalism endure, whereas the intergovernmental system's practical operations have adapted to changing political and economic conditions. The following discussion distinguishes three enduring principles from their practical operations. This comparison aims at separating principles from historical circumstances to capture the salience of federalism and the meaning of the intergovernmental system's current operations.

Federalism Is Multiple, Not Unitary

A fundamental principle of federalism is its multiple governments.[2] A citizen in a federal country belongs to two jurisdictions. This principle defines federalism, distinguishing it from nations with conventional, unitary sovereignty. Although federalism has an ancient heritage, the principle is unusual and sophisticated because it contradicts governments' tradition of unitary rule.

The core concept of multiple government contradicts the model of simple, single government: integrated bureaucracy. As a principle, multiplicity disdains the premise of bureaucracy: certainty. Specialization has gradually fused the U.S. system's multiple governments into sectors. This process of forging sectors from many agencies obscures the critical contradiction between the principles of multiple governments and bureaucratic rationality. The contradiction is left unresolved, but elaborates itself in sectoral dominance underscoring the "one best way" of premature programming. The constant interactions and episodic shifts within sectors testify to the continuing reality of multiple governments.

Multiple governments also permit and encourage government to vary its activities, allowing combinations of responsibilities for different activities at different times and in different places. The national government might mint coins, whereas the state might regulate when gin can be served, and they collaborate on canals. In another era, states could start subsidized housing programs, whereas the national government encouraged land-grant colleges. Federalism's ability to vary met the demands of an expanding political economy with many different combinations of governmental actions. Today, sectoral dominance constrains the intergovernmental system's ability to vary activities.

Federalism Is Based on Checks

The U.S. Constitution is based on checks and balances. Their purpose is to protect against tyrants and tyrannical majorities. The national level of government was intended to check the tyranny of local factions (Madison et al., 1937/1787-1788, No. 10).

Conclusions

Two hundred years of development have reversed these territorial checks. Vertically linked sectors have emerged as powerful factions. Today's primary threat to democracy is not a despot or local majorities, but sectoral dominance. In a reversal of *The Federalist No. 10*, a sector is "united and activated by some common . . . interest, adverse to the rights of other citizens or to the permanent and aggregate interests of the community (Madison et al., 1937/1787-1788).

Instead of one large national level neatly checking[3] many local factions, thousands of dispersed area governments must check disjointed sectoral impositions. The messy complexity of many unconnected jabs of resistance confounds the neat, reciprocal balance of checks in federalism's original design. Collectively as well as individually, area governments are weak checks on powerful sectors.

Federalism Is Conservative

Checks, key elements in federalism's design, underscore its basic conservatism. Federalism protects against doing harm at the expense of facilitating doing good. It protects minorities at the expense of aiding majorities. As a design against tyranny, federalism often seems a design against government effectiveness. Federalism is cautious, discouraging extreme changes.

Some argue that federalism is doubly conservative because it was designed to entrench the elite (Beard, 1913). Today's federalism is conservative in the second derogatory sense: It reinforces the establishment. By serving sectors' vested interests, the intergovernmental system has entrenched many privileged specialties. Today's elite extends beyond the pocketful of educated plantation owners the constitution may have originally protected. Now the privileged include not only agribusiness but a range of vested, special interests, such as everyday physicians, members of the American Medical Association. Some of the special interests (e.g., elementary school teachers) would never call themselves privileged.

Today's intergovernmental system generates a third sense of conservatism. The only questions sectoral dominance permits are sector serving. Typically, a sector asks itself, "How can we achieve our goal better, faster, or cheaper?" For example, the crime sector asks, "How should we reduce our assault and battery crime rate?" If officials were pushed to soul searching, they could perhaps ask, "How shall we improve law enforcement?"

The U.S. intergovernmental structure makes it difficult to ask questions that go beyond a sector's boundaries. Such questions might include what kind of society do we want for ourselves? For the future? How do we want to divide our resources? Our responsibilities? What is right for here and now? In the rare event such a question is raised, system operations and political norms curb public deliberation to platitudes and sound bites.

CONCLUSIONS: THE TERRITORIAL BASIS OF FEDERALISM IS WEAK; LONG LIVE FEDERALISM

Assessing today's intergovernmental system against the prevailing models and basic principles of federalism shows that although territorial divisions have lost their power, the idea of federalism remains vital. Federalism created a capacity for change, for coping with uncertainty. This capacity permitted the system to respond to challenges, opportunities, and social diversity unimaginable to the founding fathers.

This flexible capacity to create connections permitted specialization and development and thus served the nation well in the past. Ironically, however, the system evolved into one with governments fused in integrated sectors, which now have vitiated territorial checks. A former capacity to cope with uncertainty thus resulted in a system that promotes delusions of certainty.

Despite these constraints, some features of federalism persist. Although territorial divisions have lost their power, multiple governments nonetheless enrich and complicate policy development and resist centralization. Federalism's complexity keeps the intergovernmental system dynamic and interactive.

POLICY IMPLICATIONS: TOWARD A NEW POLICY DEBATE AND NEW POLICIES

Current policy debates on federalism seem trapped in the past. By fixing on territory as the key to federalism, debates have skirted important

Conclusions

issues and lost opportunities. Efforts to reduce government by devolving federal authorities to the states bypass the principle of multiple government. Other debates have pointed to intense interactions among governmental levels in a particular sector as evidence of federalism's continuing health, while ignoring the overall pattern of sectoral dominance. Still other debates have built on popular mistrust of government, arguing that simple, local-scale community programs are the real import of federalism. More constructive debates can address current actual conditions.

Three alternative policy courses can address the disjuncture between traditional territorially based federalism and today's conditions: abandon federalism, restore territorial federalism, or adapt the concept of federalism to current and coming conditions.

Popular support for the ideology of federalism makes forthright rejection of federalism politically impossible. Nobody is making this case. On the other hand, the quiet withering away of federalism is not only feasible, but already under way.

Proponents for traditional federalism rest their case on popular mistrust of central government. Restoring dual, separate federalism contradicts current socioeconomic conditions and the essence of federalism, however. Proposals to simplify, do away with government, and devolve responsibilities to state governments in the name of federalism misconstrue federalism.

If carried out, these policies would collapse the intergovernmental system's multiplicity, checks, and capacity to vary into a set of 50 disjointed unitary rules. Such proposals would sabotage the economy. At the same time, they would increase bureaucracies and economic and racial differences (Brinkley, 1996).

Less extreme efforts to counteract the trend toward national policy formulation have neither simplified nor reduced the intergovernmental system. Instead, a variety of reforms has resulted in more intergovernmental participation, once termed *hypergovernmentalization* (Lovell, 1983). A stream of presidential promises to "get government off the backs of people" and actual reforms have failed to restore traditional territorial federalism. Sectors' persistence in the face of this political opposition testifies to their power.

The case for adapting federalism to current conditions goes beyond a compromise between a recognition that territorial federalism is over and a determination to preserve the past. The case for adaptation argues that key features of federalism could equip the nation for future uncertainties. From this perspective, federalism's value lies in its original checks and ability to vary. The checks are conservative, protecting against abuse of power and inducing caution. In another way, however, checks offer the potential not for preserving the past but for confronting the future. When checks are built into the fabric of government, they permit government to face uncertainty directly because they encourage challenge. Effective checks can prompt not only the caution that grows from deliberation and debate but also consideration of much wider, daring, and potentially responsive proposals.

Because of the atrophy of territorial checks, the system no longer has checks built into its structure. Until a new structural check can be devised, policymakers must build checks and the ability to vary into planning processes, public policies, and public organizations.

By inviting challenge and skepticism, by encouraging participation and policy reformulation, government policy can not only acknowledge but use popular distrust. Shedding delusions of certainty and exploiting diversity can assert federalism's legacy and keep it poised for the future.

NOTES

1. The U.S. Supreme Court sent mixed messages on cosovereignty. It has refused to protect state cosovereignty as a constitutional provision (*Garcia v. San Antonio Metropolitan Transit Authority*, 1985). In June 1997, however, Justice Scalia's opinion for a 5-to-4 majority in *Printz v. United States*, No. 95-1478, said a provision violated "the very principle of separate state sovereignty," which he called "one of the constitution's structural protections of liberty." Congress may not require states to help administer federal programs, the Court said.

2. Some see the principle of multiple governments in terms of redundancy, providing safeguards (Landau, 1969), competition (Dye, 1990; Kincaid, 1988; Wildavsky, 1980), and multiple representation (Beer, 1978, pp. 9-21). All these meanings relate to diversity and options.

3. Collectively, the national government serves as a check against other sources of concentrated power, such as banks, media giants, and huge corporations (Brinkley, 1996).

CHAPTER 9

CONCLUSIONS FOR DECISION-MAKING PRACTICE

This book argues for intergovernmental decision making to respond to uncertainty. Chapter 8, "Conclusions for the Intergovernmental System," ends by proposing that policymakers build checks into programs to embed caution in the intergovernmental system. This chapter suggests ways to address uncertainty and embed caution into everyday planning and management practice. Previous chapters have recommended different forms of policy, planning, and organization appropriate to each set of problem conditions. This chapter recommends everyday practices that create a basis for and facilitate adopting these contingent approaches to uncertainty.

The general approach simply reverses customary practices of premature programming: deinstitutionalize, sow doubt, scrutinize feedback. The process generates complexity at the beginning rather than suppressing it until later, after a premature program has inflicted harms and provoked more cumbersome, entrenched, politically tough problems. Like a good environmental impact analysis, the proposed process anticipates issues and exposes conflicts prior to action. It encourages flexibility, sensitivity, and responsiveness.

The suggestions work at every level of government. They particularly address local practitioners, such as generalist planners in a city or county planning department or office of community development, or city managers. Most of these suggestions apply to those working at other levels of government and in sectors as well.

1. Create an organizational climate, framework, and procedures for addressing uncertainty.
   - Encourage organizational learning (Benveniste, 1989; Michael, 1971).
   - Encourage doubt and skepticism (Weick, 1979).
   - Reward problem finding.
   - Set up releases from regular duties to allow time and freedom to critique the status quo and develop alternatives.
   - Build mini-evaluations and client feedback into activities.
   - Use stakeholder analysis (Christensen, 1993) to expand understandings of consumers.
   - Conduct research for stakeholders (e.g., other city departments).
   - Infuse flexibility into the organization by relaxing rules and building slack.
   - Develop contingent performance standards that respond to problem conditions (e.g., more effective searches for technologies) for personnel and department reviews.

2. Diagnose problem conditions. Problem diagnosis (which set of problem conditions pertains) depends on who is asked, so ask widely.
   - Develop many ways to get community feedback: community hot line (phone), website, e-mail, community television with telephone feedback, as well as neighborhood meetings.
   - Examine assumptions embedded in the program and its history.
   - If all is quiet and there is no debate, the program should be in the Known Technology, Agreed Goal box. When experiments get standard, good results and staff treat these results as routine, it is time to move here.
   - If the program is not working, working erratically, or with diverse effects, it should be in the No Known Technology, Agreed Goal box. Complaints of implementation problems could be symptoms of unproven technology. Political support for the technology may have slipped, or perhaps the old technology never worked.

Conclusions for Decision Making 163

- If there is conflict but it is clear what different groups want, the program should be in the Known Technology, No Agreed Goal box. If people easily identify others who think opposite on the issue, it belongs here.
- If there are vague complaints, such as "not responsive," "city doesn't care," "we don't have a handle on this," the program should be in the Unknown Technology, No Agreed Goal box. If evaluation yields ambiguous results or there are conflicts without proposals (no agenda), then the program should be here.

3. Deter sectors from dominating decision making.
   - Invite nonexperts and outsiders of all types—from other levels of government, from other fields—to comment on everyday routines, as well as new proposals.
   - Find people who think opposite by asking one "who would have an opposite view or opinion from yours?" Then ask the same of the next person, and so on (Guba & Lincoln, 1989).
   - Use consumer focus groups to identify problems in sector proposals.
   - Use the League of Cities, the county supervisors association, and congressional representatives to avoid unfunded mandates and excessive conditions on grants in aid.
   - Use local sector members to get early information as programs are developing to shape the new program in the form of pilots and experiments.
   - Work with sector agencies to find and expand opportunities for flexibility and program designs appropriate to problem conditions.
   - Provide information useful to the sectors to be invited inside the sectors.
   - Work with the city council to have contingent performance standards apply to personnel and departments in sectors.
   - Develop communication and negotiation skills, because moving away from certainty and familiar sector norms increases uncertainty and unfamiliarity in language and motives.
   - Form task forces of different people and agencies for each issue to ensure the response is issue driven, not expert driven.
   - When a pilot program is a howling success, resist programming. Instead, develop an experimental test (e.g., different types of neighborhoods, nonprofits).
   - Label the experiment as issue-response to confer legitimacy and avoid the appearance of temporizing. Create credible alternatives to premature programming.

Box 9.1

A Day in the Life of a Planner Compared With a Day in the Life of a Planner Prepared for Uncertainty

	Usual Planner	Planner Prepared for Uncertainty
9:00	Coffee, visit with fellow planners, skim "Planning"	Coffee, check community line phone mail, faxes, Web, and e-mail for feedback
10:00	Staff meeting on progress on area plans, preparation for planning commission meeting	Staff meeting with representatives from other city departments to identify how planning can help (and learn about potential new developments in sectors)
11:00	GIS analysis of zoning changes	Conduct mini-evaluation of neighborhood pilot project with neighborhood residents and community college class
1:00	Luncheon meeting of Northern California Planners' Association, presentation in special assessment districts	Luncheon meeting with staff from neighboring cities and League of Cities to devise strategy for removing restrictive conditions on new block grant
2:00	Cope with crisis over developer backing out of development	Help transportation planners design a passenger-responsive paratransit service that includes feedback from passengers, drivers, and service agencies
3:00		Prepare for focus group of neighbors near proposed new developments
4:00	Review demographic and income data for another proposed project	Run focus group
5:00	Prepare written and graphic materials for neighborhood meeting	Analyze focus group findings
Evening	Present draft area plan to neighborhood association	Facilitate nominal group process to develop neighborhood's goals for area plan

These suggestions create preconditions and a framework for following the recommended forms of policy, organization, and planning rules appropriate for different conditions of uncertainty set forth in Chapter 8, "Conclusions for the Intergovernmental System." They are summarized in Figure 9.1.

Conclusions for Decision Making

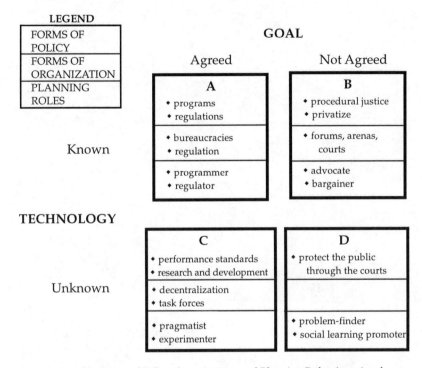

Figure 9.1. Variable Forms of Policy, Organization, and Planning Rules Associated With Variable Problem Conditions

Do not worry that following these recommendations will throw the organization into chaos. Some chaos is a good thing (Kiel, 1994). Moreover, inertia and old norms predispose organizations toward premature consensus and premature programs. Following these suggestions will help an organization open up and sow doubt and caution.

Another predictable objection: These suggestions may appear too time-consuming. But these cautious activities will save time, money, pain, and political havoc by treating actual problem conditions. Problem conditions will come to light sooner, permitting them to be treated before they develop into crisis proportions. Furthermore, the strategies for dealing with uncertainties offer many ways of dealing with problems. Eventually, as Covey (1990) recommends, with fewer fires to put out, decision makers will be able to spend more time on matters that are not urgent, but important.

REFERENCES

Adler, S. (1990). Environmental movement politics, mandates to plan, and professional planners: The dialectics of discretion in planning practice. *Journal of Architectural and Planning Research, 7*(4), 315-329.
Advisory Commission on Intergovernmental Relations. (1961). *Periodic congressional assessment of federal grants in aid for state and local governments*. Washington, DC: Author.
Advisory Commission on Intergovernmental Relations. (1995). *Characteristics of federal grant-in-aid programs to state and local governments: Grants funded FY 1995*. Information report, M-195.
Agranoff, R. J., & Rinkle, V. (1986). *Intergovernmental management: Human services problem-solving in six metropolitan areas*. Albany: State University of New York Press.
Ahlbrandt, R. S. Jr., & Brophy, P. C. (1975). *Neighborhood revitalization*. Lexington, MA: Lexington Books.
Alexander, E. R. (1986). After rationality, what? A review of responses to paradigm breakdown. *Journal of the American Planning Association, 50*(1), 62-69.
Alexander, E. R. (1996). After rationality: Towards a contingency theory of planning. In S. Mandelbaum, L. Mazza, & R. Burchell (Eds.), *Explorations in planning theory*. New Brunswick, NJ: Center for Urban Policy Research.
Alford, R. R. (1975). *Health care politics, ideological and interest group barriers to reform*. Chicago: University of Chicago Press.
Archibald, K. (1970). *Three views of the experts' role in policy-making*. Santa Monica, CA: RAND.
Argyris, C. (1964). *Integrating the individual and the organization*. New York: John Wiley.
Bardach, E. (1977). *The implementation game*. Cambridge: MIT Press.

Bardach, E., & Lesser, C. (1996). Accountability in human services collaboratives—For what? And to whom? (Symposium on the hollow state: capacity, control, and performance in interorganizational settings). *Journal of Public Administration Research and Theory, 6*(2), 197.

Barnard, C. I. (1958). *Functions of the executive.* Cambridge, MA: Harvard University Press. (Original work published 1938)

Beard, C. (1913). *An economic interpretation of the constitution.* New York: Macmillan.

Beer, S. H. (1976). The adoption of general revenue sharing: A case study in public sector politics. *Public Policy, 24,* 166-171.

Beer, S. H. (1977). A political scientist's view of fiscal federalism. In W. Oates (Ed.), *The political economics of fiscal federalism* (pp. 9-12). Lexington, MA: Lexington Books.

Beer, S. H. (1978). Federalism, nationalism and democracy in America. *American Political Science Review, 72*(1), 9-21.

Bendor, J. B. (1985). *Parallel systems: Redundancy in government.* Berkeley: University of California Press.

Benson, J. K. (1982). A framework for policy analysis. In D. L. Rogers & D. A. Whettern (Eds.), *Interorganizational coordination* (pp. 137-176). Ames: Iowa State University Press.

Beneviste, G. (1989). *Mastering the politics of planning: Crafting credible plans and policies that make a difference.* San Francisco: Jossey-Bass.

Berger, P. L., & Luckmann, T. (1966). *The social construction of reality: A treatise in the sociology of knowledge.* Garden City, NY: Doubleday.

Bingham, G. (1986). The growth of the environmental dispute resolution field. In G. Bingham (Ed.), *Resolving environmental disputes: A decade of experience.* Washington, DC: Conservation Foundation.

Blanco, H. (1994). *How to think about social problems: American pragmatism and the idea of planning.* Westport, CT: Greenwood.

Bledstein, B. J. (1978). *The culture of professionalism.* New York: Norton.

Boehm, M. H. (1931). Federalism. In E. R. A. Selgiman (Ed.), *Encyclopaedia of the social sciences.* New York: Macmillan.

Bolan, R. (1969). Community decision behavior: Culture of planning. *Journal of the American Institute of Planners, 35*(5), 301-310.

Bollens, J. (1961). *Special district government of the United States.* Berkeley: University of California Press.

Bollens, S. A. (1992). State growth management: Intergovernmental frameworks and policy objectives. *Journal of the American Planning Association, 58*(4), 454-467.

Boyte, H. (1990). The growth of citizen politics: Stages in local community organizing. *Dissent, 37,* 513-518.

Brinkley, A. (1996, August 18). Big government is a check. *New York Times Magazine,* p. 37.

Bryson, J. M., & Crosby, B. C. (1992). *Leadership for the common good: Tackling public problems in a shared-power world.* San Francisco: Jossey-Bass.

Bryson, J. M., & Delbecq, A. L. (1979). A contingent approach to strategy and tactics in project planning. *Journal of the American Planning Association, 45*(2), 167-179.

Buchanan, J., & Tullock, G. (1962). *The calculus of consent.* Ann Arbor: University of Michigan Press.

Burstein, P. (1981). Policy domains: Organization, culture, and policy outcomes. *Annual Review of Sociology, 17,* 327-350.

CALFED Bay-Delta Program. (1996). *Phase I progress report: CALFED Bay Delta Program.* Sacramento, CA: Author.

Cartwright, T. J. (1973). Problems, solutions, strategies. *Journal of the American Institute of Planners, 39*(3), 179-187.

References

Checkoway, B. (Ed.). (1993). Paul Davidoff and advocacy planning in retrospect. *Journal of the American Planning Association, 60*(2), 139-161.

Chisholm, D. (1989). *Coordination without hierarchy: Informal structures in multiorganizational systems.* Berkeley: University of California Press.

Christensen, K. (1993). Teaching savvy. *Journal of Planning Education and Research, 12,* 202-212.

Christensen, K., & Teitz, M. (1980). The housing assistance plan: Promise and reality. In D. Marshall & R. Montgomery (Eds.), *Housing in the 1980s.* Lexington, MA: Lexington Books.

Corwin, E. S. (1934). *The twilight of the supreme court.* Princeton, NJ: Princeton University Press.

Covey, S. R. (1990). *The seven habits of highly effective people.* New York: Fireside.

Cyert, R., & March, J. (1963). *A behavioral theory of the firm.* Englewood Cliffs, NJ: Prentice Hall.

Dahl, R. (1961). *Who Governs?* New Haven, CT: Yale University Press.

Dalton, L. C. (1986). Why the rational paradigm persists: The resistance of professional education and practice to alternative forms of planning. *Journal of Planning Education and Research, 5,* 147-153.

Davidoff, P. (1965). Advocacy and pluralism in planning. *Journal of the American Institute of Planners, 31*(4), 331-338.

de Neufville, J. I., & Christensen, K. S. (1981). Is optimizing really best? In D. J. Palumbo, S. B. Fawcett, & P. Wright (Eds.), *Evaluating and optimizing public policy.* Lexington, MA: Lexington Books.

De Parles, J. (1997). Getting opal cables to work. *New York Times Magazine,* August 24, p. 32.

Derthick, M. (1970). *The influence of federal grants, public assistance in Massachusetts.* Cambridge, MA: Harvard University Press.

DiMaggio, P. J., & Powell, W. W. (1983). The iron cage revisited: Institutional isomorphism and collective rationality in organizational fields. *American Sociological Review, 48,* 147-160.

Donahue, J. D. (1989). *The privatization decision: Public ends, private means.* New York: Basic Books.

Douglas, M. (1986). *How institutions think.* Syracuse, NY: Syracuse University Press.

Downs, A. (1966). *Inside bureaucracy.* Boston: Little, Brown.

Dror, Y. (1968). *Public policymaking re-examined.* San Francisco: Chandler.

Dye, T. R. (1990). *American federalism: Competition among governments.* Lexington, MA: D.C.Heath.

Elazar, D. J. (1962). *The American partnership.* Chicago: University of Chicago Press.

Elazar, D. J. (1968). Federalism. In *Encyclopaedia of social sciences.* New York: Macmillan.

Elazar, D. J. (1991). Cooperative federalism. In D. Kenyon & J. Kincaid (Eds.), *Competition among states and local governments: Efficiency and equity in American federalism.* Washington, DC: Urban Institute Press.

Forester, J. (1983). Planning in the face of power. *Journal of the American Planning Association, 48*(1), 67-80.

Forester, J. (1989). *Planning in the face of power.* Berkeley: University of California Press.

Forester, J. (1993). *Critical theory, public policy, and planning practice: Toward a critical pragmatism.* Albany, NY: SUNY Press.

Forester, J. W. (1969). *Urban dynamics.* Cambridge: MIT Press.

Friedmann, J. (1973). *Retracking America: A theory of transactive planning.* Garden City, NJ: Anchor Press/Doubleday.

Friedmann, J. (1987). *Knowledge and action: Mapping the planning theory terrain.* Princeton, NJ: Princeton University Press.

Gale, D. E. (1992). Eight state-sponsored growth management programs: A comparative analysis. *Journal of the American Planning Association, 58*(4), 425-440.

Gans, H. (1962). *The urban villagers.* New York: Free Press.

Garcia v. San Antonio Metropolitan Transit Authority, 105 S.Ct. 1005 (1985).

General Accounting Office. (1993). *Natural resources restoration: Use of Exxon Valdez oil spill settlement funds* (Briefing report to the chairman, Committee on Natural Resources, House of Representatives, GAO/RCED-93-206BR). Washington, DC: Author.

Gerth, H. H., & Mills, C. W. (1946). *From Max Weber.* New York: Oxford University Press.

Godschalk, D. R. (1992). Negotiating intergovernmental development policy conflicts: Practice-based guidelines. *Journal of the American Planning Association, 58*(3), 368-378.

Grodzins, M. (1966). *The American system.* Chicago: Rand McNally.

Gruber, J. (1994). *Coordinating growth management through consensus building: Incentives and the generation of social, intellectual and political capital* (Working Paper No. 617). Berkeley: University of California, Institute of Urban and Regional Development.

Guba, E. G., & Lincoln, Y. S. (1989). *Fourth generation evaluation.* Newbury Park, CA: Sage.

Heclo, H. (1977). *A government of strangers: Executive politics in Washington.* Washington, DC: Brookings Institution.

Hirschmann, A. O., & Lindblom, C. (1962). Economic development, research and development and policy making: Some converging views. *Behavioral Science, 7*(2), 211-222.

Hjern, B., & Porter, D. O. (1981). Implementation structures: A new unit for administrative analysis. *Organizational Studies, 2*(3), 211-237.

Hoch, C. (1994). *What planners do: Power, politics, and persuasion.* Chicago: American Planning Association.

Hoffman, S. (1959). A real division of powers in the writings of French political writers. In A. Maass (Ed.), *Area and power.* Glencoe, IL: Free Press.

Howitt, A. M. (1984). *Managing federalism: Studies in intergovernmental relations.* Washington, DC: CQ Press.

Huntington, S. (1959). The founding fathers and the division of powers. In A. Maass (Ed.), *Area and power.* Glencoe, IL: Free Press.

Innes, J. E. (1992). Group processes and the social construction of growth management: Florida, Vermont, and New Jersey. *Journal of the American Planning Association, 58*(4), 440-454.

Innes, J. E. (1995). Planning theory's emerging paradigm: Communicative action and interactive practice. *Journal of Planning Education and Research, 14*(3), 183-189.

Innes, J. E. (1996). Planning through consensus building: A new view of the comprehensive planning ideal. *Journal of the American Planning Association, 62*(4), 460-473.

Innes, J. E., Gruber, J., Neuman, M., & Thompson, R. (1994). *Coordinating growth and environmental management through consensus building.* Berkeley: University of California at Berkeley, California Policy Seminar.

Innovative swap of air pollution credits. (1994, November 19). *San Francisco Chronicle*, p. A4.

Kaplan, A. (1961). *The new world of philosophy.* New York: Random House.

Kaufman, H. (1977). *Red tape: Its origins, uses, and abuses.* Washington, DC: Brookings Institution.

Kenyon, D., & Kincaid, J. (Eds.). (1991). *Competition among states and local governments: Efficiency and equity in American federalism.* Washington, DC: Urban Institute Press.

References

Kestnbaum, M. (Chair). (1955). *Commission on intergovernmental relations: A report to the president for transmittal to the 84th Congress* (First Session, House Document No. 198). Washington, DC: Government Printing Office.

Kettl, D. F. (1988). *Government by proxy*. Washington, DC: CQ Press.

Kiel, L. D. (1994). *Managing chaos and complexity in government: A new paradigm for managing change, innovation, and organizational renewal*. San Francisco: Jossey-Bass.

Kincaid, J. (1991). The competitive challenge to cooperative federalism: A theory of federal democracy. In D. Kenyon & J. Kincaid (Eds.), *Competition among states and local governments: Efficiency and equity in American federalism*. Washington, DC: Urban Institute Press.

Kingdon, J. W. (1984). *Agendas, alternatives, and public policies*. New York: HarperCollins.

Kunioka, T., & Rothenberg, L. S. (1993). The politics of bureaucratic competition: The case of natural resource policy. *Journal of Policy Analysis and Management, 12*, 700-725.

Ladd, E. C. (Ed.). (1995). *America at the polls*. Storrs: University of Connecticut, Roper Center.

Landau, M. (1969, July/August). Redundancy, rationality and the problem of duplication and overlap. *Public Administration Review, 29*(3), 346-358.

Landau, M. (1973). Federalism, redundancy, and system reliability. *Publius, 3*(2), 173-196.

Landau, M., & Stout, R., Jr. (1979). To manage is not to control. *Public Administration Review, 33*(6), 533-542.

La Porte, T. R. (1975). Complexity and uncertainty: Challenge to action. In T. R. La Porte (Ed.), *Organized social complexity: Challenge to politics and policy* (pp. 332-356). Princeton, NJ: Princeton University Press.

Lawless, M. W. (1981). Directed interorganizational systems: Network strategy-making in public service delivery. Unpublished manuscript, California State University, Northridge.

Lawrence, P. R., & Lorsch, J. W. (1967). *Organization and environment*. Boston: Harvard Business School.

Levine, R. A. (1972). *Public planning: Failure and redirection*. New York: Basic Books.

Lewis, J. B. (Ed.). (1967). *Antifederalist vs. federalist papers*. San Francisco: Chandler.

Lindblom, C. E. (1959). The science of muddling through. *Public Administration Review, 19*(2), 79-88.

Lindblom, C. E. (1965). *The intelligence of democracy: Decision-making through mutual adjustment*. New York: Free Press.

Lovell, C. H. (1983). Some thoughts on hyperintergovernmentalization. In R. H. Leach (Ed.), *Intergovernmental relations in the 1980s* (pp. 87-97). New York: Marcel Dekker.

Lowi, T. (1969). *The end of liberalism*. New York: Norton.

Madison, J., Hamilton, A., & Jay, J. (1937). *The federalist*. New York: Random House. (Original work published 1787-1788)

Mandell, M. P. (1990). Network management: Strategic behavior in the public sector. In R. W. Gage & M. P. Mandell (Eds.), *Strategies for managing intergovernmental policies and networks*. New York: Praeger.

March, J., & Olsen, J. P. (1976). *Ambiguity and choice in organizations*. Bergen: Universitetforlaget.

Markusen, A. (1976). Class and urban social expenditures: A local theory of the state. *Kapitalistate, 4*(5), 51-65.

Marris, P. (1984). Social change and reintegration. *Journal of Planning Education and Research, 2*(1), 54-61.

Marris, P. (1996). *The politics of uncertainty*. New York: Routledge.

Mayo, E., et al. (1951). The Western Electric studies. In S. D. Hoslett (Ed.), *Human factors of management*. New York: Harper.

McConnell, G. (1966). *Private power and American democracy.* New York: Knopf.
McCulloch v. Maryland, 4 Wheat, 316, 431 (1819).
McLean, J. (1952). *Politics is what you make it* (Public Affairs Pamphlet No. 181). New York: Public Affairs Committee, Inc.
Merton, R. (1957). *Social theory and social structure* (rev. ed.). New York: Free Press.
Meyer, J. W., & Rowan, B. (1977). Institutionalized organizations: Formal structure as myth and ceremony. *American Journal of Sociology, 83,* 340-363.
Michael, D. (1971). *On learning to plan and planning to learn.* San Francisco: Jossey-Bass.
Miller, G. J. (1981). *Cities by contract: The politics of municipal incorporation.* Cambridge: MIT Press.
Milliken v. Bradley, 418 U.S. 717 (1974).
Mills, C. W. (1956). *The power elite.* New York: Oxford University Press.
Milward, H. B. (1982). Interorganizational policy systems and research on public organizations. *Administration and Society, 13*(4), 457-478.
Moe, T. (1989). The politics of bureaucratic structure. In J. E. Chubb & P. E. Person (Eds.), *Can the government govern?* Washington, DC: Brookings Institution.
Moore, T. (1978). Why allow planners to do what they do? *Journal of the American Institute of Planners, 44*(4), 387-398.
Morris, M. (1997). Taking the bite out of big projects. *Planning, 63*(2), 20-24.
Muller, B. (1994). The rise and fall of the entrepreneurial state: The case of Texas in the 1980s. *Berkeley Planning Journal, 9,* 1-21.
National Commission on Urban Problems. (1969). *Building the American city.* New York: Praeger.
Niskamin, W. A. (1979). Competition among government bureaus. *American Behavioral Scientist, 22,* 517-524.
North American Waterfowl Management Plan. (1995). *1994 annual report.* Washington, DC: Government Printing Office.
Okun, A. M. (1975). *Equality and efficiency: The big tradeoff.* Washington, DC: Brookings Institution.
Osborne, D., & Gaebler, T. (1992). *Reinventing government: How the entrepreneurial spirit is transforming the public sector.* New York: Penguin.
Ostrom, E. (1990). *Governing the commons: The evolution of institutions for collective action.* Cambridge, MA: Press Syndicate of the University of Cambridge.
Ostrom, V. (1974). *The intellectual crisis in American public administration* (rev. ed.). Tuscaloosa: University of Alabama Press.
Ostrom, V. (1977). Some problems in doing political theory: A response to Golembiewski's "Critique." *American Political Science Review, 71*(4), 1508-1525.
Ostrom, V., Tiebout, C., & Warren, R. (1961). The organization of government in metropolitan areas: A theoretical inquiry. *American Political Science Review, 55*(4), 831-842.
O'Toole, L. (1989). Alternative mechanisms for multiorganizational implementation. *Administration and Society, 21,* 313-339.
O'Toole, L. (1993). Interorganizational policy studies: Lessons drawn from implementation research. *Journal of Public Administration Research and Theory, 3*(2), 232-251.
Pamuk, A., & Christensen, K. (1989). Preliminary findings on San Francisco Bay Area nonprofit housing developers. *Berkeley Planning Journal, 4,* 19-36.
Parsons, T. (1960). *Structure and process in modern society.* Glencoe: Free Press.
Perin, C. (1977). *Everything in its place.* Princeton. NJ: Princeton University Press.
Perrow, C. (1970). *Organizational analysis: A sociological view.* Belmont, CA: Wadsworth.
Peterson, P. E., Rabe, B. G., & Wong, K. K. (1986). *When federalism works.* Washington, DC: Brookings Institution.

References

Pfeffer, J. (1972). Merger as a response to organizational interdependence. *Administrative Science Quarterly, 17,* 382-394.

Pfeffer, J., & Salancik, G. (1978). *The external control of organizations: A resource dependence perspective.* New York: Harper & Row.

Popper, F. (1987). The environmentalist and the lulu. In R. W. Lake (Ed.), *Resolving locational conflict.* New Brunswick, NJ: Center for Urban Policy Research.

Popper, K. (1959). *The logic of scientific discovery.* New York: Basic Books.

Powell, W. W. (1990). Neither market nor hierarchy: Network forms of organizations. In B. M. Staw & L. L. Cummings (Eds.), *Research in organizational behavior,* 295-336. Greenwich, CT: JAI.

Powell, W. W., & DiMaggio, P. (Eds.). (1991). *The new institutionalism in organizational analysis.* Chicago: University of Chicago Press.

Printz v. United States, No. 95-1478 (1997).

Provan, K. G. (1983). The federation as an interorganizational linkage network. *Academy of Management Review 8*(61), 78-89.

Raelin, J. A. (1980). A mandated basis of interorganizational relations: The legal-political network. *Human Relations, 33*(1), 57-68.

RAND Corporation. (1982). *Experimental housing allowance program evaluation.* Santa Monica, CA: Author.

Riker, W. H. (1964). *Federalism: Origin, operations, significance.* Boston: Little, Brown.

Rittel, H., & Webber, M. M. (1973). Dilemmas in a general theory of planning. *Policy Sciences, 4*(2), 155-169.

Rivlin, A. (1971). *Systematic thinking for social action.* Washington, DC: Brookings Institution.

Rondinelli, D. (1973). Urban planning as policy analysis: Management of urban change. *Journal of the American Institute of Planners, 39*(1), 13-22.

Sanford, T. (1967). *Storm over the states.* New York: McGraw-Hill.

Sassen, S. (1991). *The global city: New York, London, Tokyo.* Princeton, NJ: Princeton University Press.

Saxenian, A. (1994). *Regional advantage: Culture and competition in Silicon Valley and Route 128.* Cambridge, MA: Harvard University Press.

Schon, D. (1983). *The reflective practitioner.* New York: Basic Books.

Schultz, C. (1977, May). The public use of private interest. *Harper's,* p. 254.

Scott, W. R. (1992). *Organizations: Rational, natural, and open systems.* Englewood Cliffs, NJ: Prentice Hall.

Seidman, H. (1970). *Politics, position and power.* New York: Oxford University Press.

Selling growth. (1995, January 22). *San Francisco Examiner,* pp. C-1, C-3.

Simon, H. (1976). *Administrative behavior* (3rd ed.). New York: Macmillan.

Steinbrunner, J. D. (1974). *Cybernetic theory of decision.* Princeton, NJ: Princeton University Press.

Stoker, R. (1991). *Reluctant partners: Implementing federal policy.* Pittsburgh, PA: University of Pittsburgh Press.

Sundquist, J. P. (1969). *Making federalism work.* Washington. DC: Brookings Institution.

Susskind, L. (1981). The importance of citizen participation and consensus building in the land use planning process. In J. Innes (Ed.), *The land use policy debate.* New York: Plenum.

Susskind, L., & Cruickshank, J. (1987). *Breaking the impasse: Consensual approaches to resolving public disputes.* New York: Basic Books.

Susskind, L., & Ozawa, C. (1983, October). Mediated negotiations in the public sector: The planner as mediator. Presented at the conference of the Association of Collegiate Schools of Planning, San Francisco.

Taylor, S. (1984). *Making bureaucracies think: The environmental impact strategy of administrative reform.* Stanford, CA: Stanford University Press.

Thomas, C. W. (1994, March). *Protecting cooperative turf: Interagency strategies for managing human impacts on native species in California.* Delivered at the annual meeting of the Western Political Science Association, Albuquerque, NM.

Thomas, C. W. (1997). *Bureaucratic landscapes: Interagency cooperation and the preservation of biodiversity.* Unpublished PhD dissertation, University of California, Berkeley.

Thompson, J. D. (1967). *Organizations in action.* New York: McGraw-Hill.

Thompson, J. D., & Tuden, A. (1959). Strategies, structures and processes of organizational decision. In J. D. Thompson et al. (Eds.), *Comparative studies in administration.* Pittsburgh, PA: University of Pittsburgh Press.

Tiebout, C. (1956). A pure theory of local expenditure. *Journal of Political Economy, 64*(5), 416-424.

Truman, D. B. (1951). *The governmental process* (2nd ed.). New York: Knopf.

Webber, M. M. (1964). Urban place and nonplace urban realm. In M. M. Webber (Ed.), *Explorations into urban structure.* Philadelphia: University of Pennsylvania Press.

Webber, M. M. (1978). A difference paradigm for planning. In R. Burchell & G. Sternly (Eds.), *Planning theory in the 1980s.* New Brunswick, NJ: Rutgers University, Center for Urban Policy Research.

Webster-Merriam. (1956). *Webster's New Collegiate Dictionary.* Springfield, MA: Merriam.

Weick, K. (1979). *Social psychology of organizing* (2nd ed.). Reading, MA: Addison-Wesley.

Wildavsky, A. (1964). *The politics of the budgetary process.* Boston, MA: Little, Brown.

Wildavsky, A. (1980). *Bare bones: The federal skeleton in the closet of American government.* Unpublished manuscript.

Williams, O. P. (1971). *Metropolitan political analysis: A social access approach.* New York: Free Press.

Wills, G. (1996, August 11). It's his party. *New York Times Magazine,* pp. 30-37, 52, 55, 57-59.

Wilson, J. Q. (1989). *Bureaucracy: What government agencies do and why they do it.* New York: Basic Books.

Wright, D. S. (1983). Managing the intergovernmental scene: The changing dramas of federalism, intergovernmental relations, and intergovernmental management. In W. Eddy (Ed.), *Handbook of organization management* (pp. 417-454). New York: Marcel Dekker.

Wright, D. S. (1978). *Understanding intergovernmental relations.* North Scituate, MA: Duxbury.

Wright, D. S. (1988). *Understanding intergovernmental relations.* Pacific Grove, CA: Brooks/Cole.

Wright, D. S. (1990). Conclusion: Federalism, intergovernmental relations, and intergovernmental management: Conceptual reflections, comparisons, and interpretations. In R. W. Gage & M. P. Mandell (Eds.), *Strategies for managing intergovernmental policies and networks.* New York: Praeger.

Yin, R. (1979, February). *Creeping federalism: The federal impact of the structure and function of local government.* Paper presented at a conference on "The Urban Impacts of Federal Policies," sponsored by the Department of Housing and Urban Development, Office of Policy Development and Research, and Johns Hopkins University, Washington, DC.

Ylvisaker, P. (1959). Reasons, means and criteria for division of powers. In A. Maass (Ed.), *Area and power.* Glencoe, IL: Free Press.

Zysman, J. (1977). *Political strategies for industrial order.* Berkeley: University of California Press.

INDEX

Area, 52-56
   See also Sectors on area

Bargaining, 106, 140, 143-144

Certainty:
   predispositons toward, 97-100
   See also Problem conditions
Cities. See Area
Collaboration, 32-39
Complexity, 2-8, 44, 62, 80-83, 147-148, 161
Conflict, 39-44

Decision making:
   interaction 56-62
   piecemeal, 85

Experimentation, 104
   constrained, 105-106

Federalism, 12 -21, 153-160.
   dual, 13-14, 17-18,
   functional, 13-14, 15-17, 19-20
   multicentered, 13, 15-16, 18-19
   principles 155-158
   policy debate 158-160.
Function. See Specialization

Goal:
   deterred debate, 84
   mega, 66-67
   multiple, conflicting, 24-25
   See also Problem conditions

Innovation. See Experimentation.
Interdependence 24-25
Intergovernmental system, 2-8, 12
   and planning, 2-8, 143-144
Intergovernmental relations, 11-12.
   See Intergovernmental system

Laissez faire, 129, 135

Market-like policies. See Public choice
Mission. See Goal

Organization form, 132-135

Planning, 2-8, 135-142, 164
   and intergovernmental system, 2-8, 143-144
   overcome constraints 145-147
Policy form, 119-132
Privatizing, 126-127
Problem conditions, 90-91

175

    and organization, 132-135
    and planning, 136-142
    and policy form, 119-132
    diagnosis 162-163.
    expectations for, 90-96
    known technology, agreed goal, 91-92, 121, 132, 136
    known technology, no agreed goal, 94, 125, 134, 139
    unknown technology, agreed goal, 92-93, 122, 133, 137,
    unknown technology, no agreed goal, 95, 129, 134, 141
Programming, 132-133
    premature, 101-103
Public choice:
    collective, 125-126
    dividing, 126-129
Public participation:
    skewed access, 84

Resource dependency, 76

Sector, 50-52
    distinct, 73
    dynamics, 66-73
    learning, 75
    proliferation, 73-75
    stable, 103-104
Sectors on areas, 76-77, 106-109
    area treatment, 79-80
    disjointed, 80-82
    dominance, 82-83, 108-109, 163
    intended/unintended difference, 77-79
Specialization, 20
    See also Sector
System:

Task division 27-31
    non-routine, 29-30
    routine 27-28
Technology, 67
    reform 69-73
    See also Problem conditions
Territory, 52-55
    specialization, 53-54
    See also Area

Uncertainty, 152-153
    climate for, 162
    over means and ends, 109-111
    See also Problem conditions

Variability, 116-118
Vertical chains, 48-50. *See* Sector

ABOUT THE AUTHOR

Karen Stromme Christensen is Assistant Professor at the Department of City and Regional Planning, University of California, Berkeley. She has been engaged in intergovernmental issues since working at the U.S. Department of Housing and Urban Development and the Federal Regional Council. These concerns prompted her to return to academic life and earn master's and PhD degrees in city planning at the University of California, Berkeley. She continues to conduct research on intergovernmental issues and affordable housing. Her current projects include examination of the institutional supports necessary for nonprofit housing developers and historical analysis of planning theory. She recently received the Chaster Rapkin Award (for outstanding article) in the *Journal of Planning Education and Research* for "Teaching Savvy," on the organizational politics of planning.